Life Sciences

Until the nineteenth century, the various subjects now known as the life sciences were regarded either as arcane studies which had little impact on ordinary daily life, or as a genteel hobby for the leisured classes. The increasing academic rigour and systematisation brought to the study of botany, zoology and other disciplines, and their adoption in university curricula, are reflected in the books reissued in this series.

Icones Plantarum

This world-famous work was begun by Sir William Jackson Hooker (1785–1865) in 1837, and the ten volumes reissued here were produced under his authorship until 1854, at which point his son, Joseph Dalton Hooker (1817–1911) continued the work of publication. Hooker's own herbarium, or collection of preserved plant specimens, was so extensive that at one point he stored it in one house and lived in another; it was left to the nation on his death. Each volume contains 100 line drawings of plants, and each is accompanied by a full Latin description, with notes in English on habitat and significant features. The order of the plants in each volume is not systematic, but two 'indexes' at the beginning provide plant lists, in alphabetical order and 'arranged according to the natural orders'.

Cambridge University Press has long been a pioneer in the reissuing of out-of-print titles from its own backlist, producing digital reprints of books that are still sought after by scholars and students but could not be reprinted economically using traditional technology. The Cambridge Library Collection extends this activity to a wider range of books which are still of importance to researchers and professionals, either for the source material they contain, or as landmarks in the history of their academic discipline.

Drawing from the world-renowned collections in the Cambridge University Library, and guided by the advice of experts in each subject area, Cambridge University Press is using state-of-the-art scanning machines in its own Printing House to capture the content of each book selected for inclusion. The files are processed to give a consistently clear, crisp image, and the books finished to the high quality standard for which the Press is recognised around the world. The latest print-on-demand technology ensures that the books will remain available indefinitely, and that orders for single or multiple copies can quickly be supplied.

The Cambridge Library Collection will bring back to life books of enduring scholarly value (including out-of-copyright works originally issued by other publishers) across a wide range of disciplines in the humanities and social sciences and in science and technology.

Icones Plantarum

*Or, Figures, with Brief Descriptive Characters
and Remarks of New or Rare Plants,
Selected from the Author's Herbarium*

VOLUME 5

WILLIAM JACKSON HOOKER

CAMBRIDGE
UNIVERSITY PRESS

CAMBRIDGE UNIVERSITY PRESS

Cambridge, New York, Melbourne, Madrid, Cape Town,
Singapore, São Paolo, Delhi, Tokyo, Mexico City

Published in the United States of America by Cambridge University Press, New York

www.cambridge.org
Information on this title: www.cambridge.org/9781108039253

© in this compilation Cambridge University Press 2011

This edition first published 1842
This digitally printed version 2011

ISBN 978-1-108-03925-3 Paperback

ICONES PLANTARUM;

OR

FIGURES,

WITH

BRIEF DESCRIPTIVE CHARACTERS AND REMARKS,

OF

NEW OR RARE PLANTS,

SELECTED FROM THE AUTHOR'S HERBARIUM.

By SIR WILLIAM JACKSON HOOKER, K.H.,

LL.D., F.R.A., AND L.S.

VICE-PRESIDENT OF THE LINNÆAN SOCIETY,

MEMBER OF THE IMP. ACAD. NAT. CUR., ETC., ETC., ETC.

HONORARY MEMBER OF THE ROYAL IRISH ACADEMY, OF THE ROYAL MEDICAL AND

CHIRURGICAL SOC. OF LONDON, ETC., ETC.

AND

DIRECTOR OF THE ROYAL BOTANIC GARDENS, KEW.

VOL. I. NEW SERIES,

OR VOL V. OF THE ENTIRE WORK.

LONDON:

HIPPOLYTE BAILLIÈRE,

FOREIGN BOOKSELLER TO THE ROYAL SOCIETY, TO THE ROYAL COLLEGE OF SURGEONS,

AND TO THE ROYAL MEDICO-CHIRURGICAL SOCIETY,

219, REGENT STREET.

PARIS: J. B. BAILLIÈRE, RUE DE L'ÉCOLE DE MÉDECINE.

MDCCCXLII.

LONDON:
PRINTED BY SCHULZE AND CO., 13, POLAND STREET.

INDEX

TO THE

PLANTS CONTAINED IN VOLUME I.,

(OR VOL. V. OF THE ENTIRE WORK;)

ARRANGED ACCORDING TO THEIR NATURAL ORDERS.

INDEX

TO THE

PLANTS CONTAINED IN VOLUME I.,

(OR VOL. V. OF THE ENTIRE WORK;)

ALPHABETICALLY ARRANGED.

TAB. CDI.

AULAYA SQUAMOSA. *Harv.*

Floribus spicatis densis, corollæ limbo concavo integerrimo.
Harv. Gen. of S. African Pl. p. 250.
Orobanche squamosa. *Thunb. Fl. Cap. p.* 455.
HAB.   Cape of Good Hope; sandy hillocks in low places;
Swartland, Saldanha Bay, Piqueberg and Verloren Valley.
Thunberg.   " The only specimens I have yet seen were
gathered at Brach-fontein by *Mrs. Van Schwon.*" *(Hon. W.
H. Harvey).*
This Mr. Harvey describes as having stems 2-3 feet high,
simple or branched, closely covered with appressed orange or
golden scales, the calyces bright orange and yellow, the tube of
the corolla a brilliant flaring yellow, and the limb deep orange.

Fig. 1. Front view of a flower with bractea. *f.* 2. Corolla.
f. 3. Pistil.   *f.* 4. Section of ovary.   *f.* 5, 6. Anthers and upper
part of the filaments :—*magnified.*

b

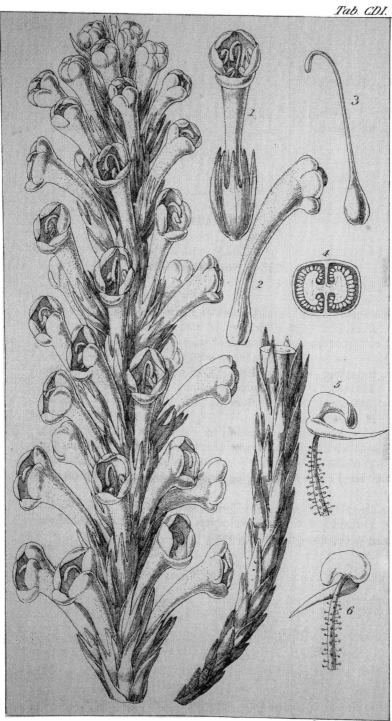

Allan & Ferguson, lithog.

TAB. CDII.

Quercus Skinneri. *Benth.*

Ramis glabris, gemmis lanatis, foliis petiolatis ovato-v. sublanceolato-oblongis sinuato-dentatis dentibus longe aristatis utrinque glabris v. subtus ad axillas venarum barbatis, fructibus sessilibus maximis, cupulæ plano-pateriformis lignosæ squamis arcte imbricatis tenuibus latis, glandula globoso-subconica lignosa basi lata umbilicata intus dissepimentis incompletis irregulariter subdivisa. *Benth. Pl. Hartw. p.* 90. *Lindl. in Gardener's Chronicle,* 1841, *p.* 116, *cum. Ic.*

Hab. Mountains, Guatemala. *G. U. Skinner, Esq.* Sides of mountains of Acatenango, Medio Monte and Quezaltenango, towards the Pacific Ocean. *Hartweg.* "Arbor pulcherrima, 50-70 pedalis. Folia utrinque viridia, iis *Q. acutifoliæ* v. *Q. Xalapensis* similia. Specimina omnia jam deflorata, florum masculorum tamen amentum unicum vidi emarcidum, generi *Quercus* omnino consimile. Glandula sæpe 2 poll. diametro, pericarpio crasso lignoso. Dissepimenta spuria ex endocarpio formata, per sulcos seminis excurrentia, valde irregularia sunt, nec loculos completos unquam efformans." *Benth. l. c.*

A figure of this curious acorn, which only yields in size to that of the following species, is given by Dr. Lindley in the Gardener's Chronicle, where he observes that the internal structure resembles that of the wallnut.

Fig. 1. Acorn : *nat. size.*

1

Allan & Ferguson, lithog.

TABS. CDIII. CDIV.

QUERCUS CORRUGATA. *n. sp.*

Ramis glabris, ramulis gemmulisque pilosis, foliis (deciduis?) petiolatis lato-lanceolatis sinuato-dentatis utrinque (etiam axillis) glabris, dentibus obtusis, cupulæ crassissimæ lignosæ brevi-turbinatæ inflexæ squamis arcte imbricatis crassis gibbosis acutis, glandula maxima sessili globoso-subconica basi latissima convexa apice depressa umbilicata umbonata. HAB. Cerro del Tamber, Guatemala, where the average temperature of the climate is 68⁰—69⁰. *G. U. Skinner, Esq.*

For the knowledge of this splendid fruited oak, which attains a height of 80 feet, we are also indebted to G. U. Skinner, Esq. The acorns are even larger than those of *Q. Skinneri,* (see our preceding plate) and the foliage and the cupula, especially, are quite different: the latter singularly rough and corrugated. Mr. Bentham observes that the cotyledons of the embryo are unequal in size and slightly uneven on the surface, but that there is nothing like the dissepiments and furrows of *Q. Skinneri,* and only a few very slightly prominent ribs on the endocarp.

Fig. 1. Acorn: *nat. size.*

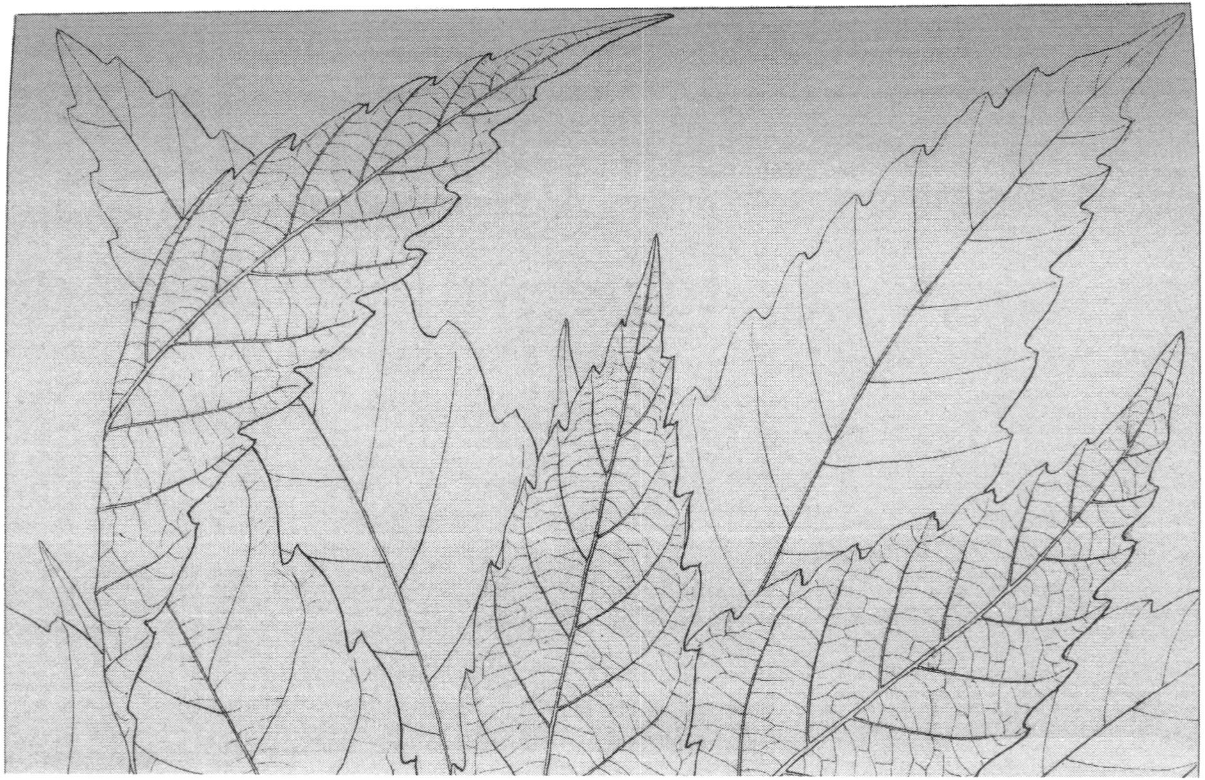

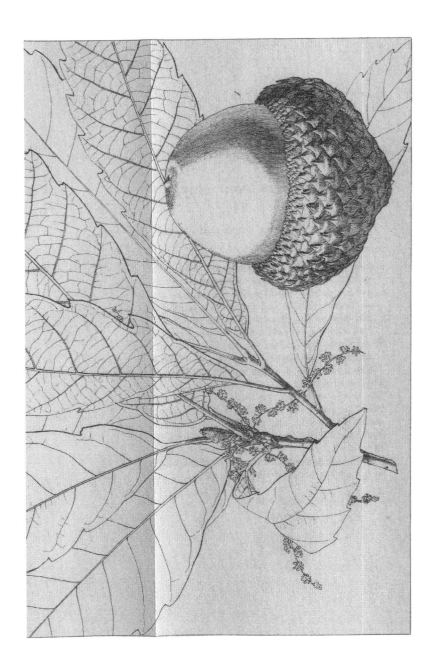

TABS. CDV. CDVI.

Eucalyptus macrocarpa. *n. sp.*

Arbor ubique farinaceo-glaucescens, foliis cordato-ellipticis brevi-acuminatis, pedunculis axillaribus solitariis brevissimis uni-floris, calycis magni crassissimi operculo conico-acuminato, capsula maxima breviter hemispherica marginata lignosa 4-5 valvi.

Hab. Guangan; Swan River Colony, Australia. *Mr. J. Drummond.*

One of the finest among the many fine plants lately sent to me by Mr. Js. Drummond from the Swan River Colony, is the present new species of *Eucalyptus.* It is noticed in Mr. Drummond's letters published in the 2d vol. of our " Journal of Botany," p. 343, and subsequent pages. Guangan is the native name of a country inland from the Swan River coast, constituting an open sandy desert, commencing about eighty miles E. S. E. of Freemantle and continuing for 200 miles. This barren sandy district is bordered by a considerable forest, composed princi-pally of two species of *Eucalyptus,* called *Urac* and *Morral* by the aborigines. The present one is the *Morral,* conspic-uous by its noble, glaucous, almost white leaves, its red flowers and its fruit, both of an unusually large size. The same species, however, Mr. Drummond has seen with white flowers.

Tab. CDV. CVI. Portion of a flowering plant : *nat. size,* and stamens : *magnified.* Tab. CDVII.

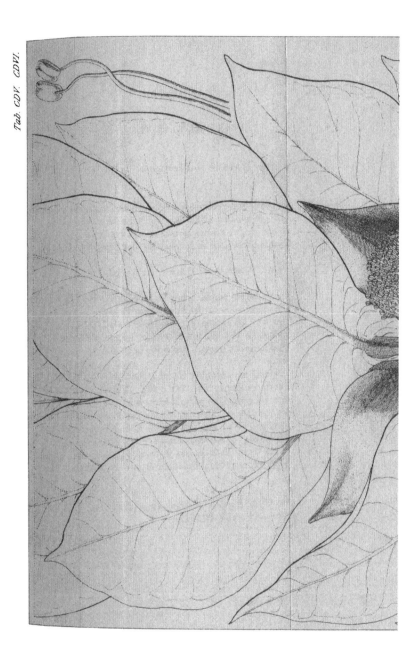

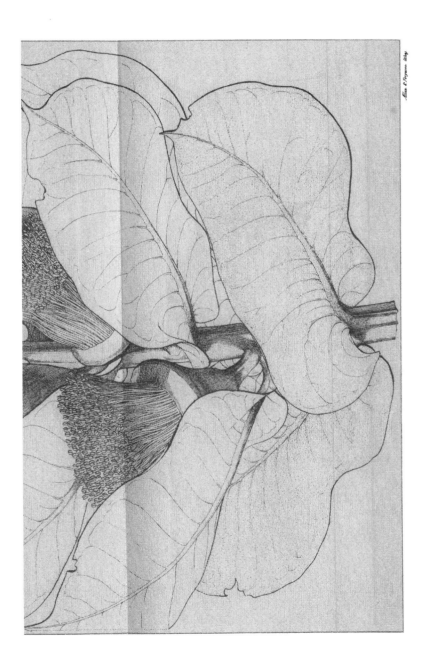

TAB. CDVII.

Eucalyptus macrocarpa. *n. sp.*

This plate represents the fruit of *Eucalyptus macrocarpa*, of which the flowering specimen is given in the preceding table.

Fig. 1. Young fruit, with 4 valves and cells, *nat. size. f.* 2. Fruit more mature, bursting into 5 valves, and containing 5 cells; *nat. size. f.* 3. Receptacle of immature seeds from *f.* 1; *nat. size. f.* 5. Immature seeds; *nat. size. f.* 6. The same seeds *magnified. f.* 4. Receptacle of seeds from *f.* 2, the seeds having fallen away; *nat. size.*

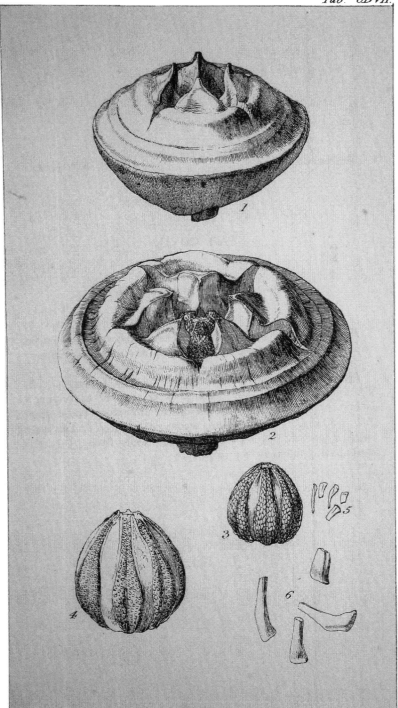

TAB. CDVIII.

MACARTHURIA. *Hügel.*

Cal. 5-sepalus, ebracteatus. *Petala* 5, oblonga, acuta, unguiculata. *Stam.* 10, fertilia, in cupulam filamentis æquilongam connata. *Antheræ* rimis lateralibus dehiscentes. *Ovarium* 3-4 -loculare, loculis 2-3-ovulatis. *Ovula* ad umbilicum strophiola crenata cincta. *Styli* loculorum numero, a basi distincti.— Frutex *ramosus, aphyllus* (?), *ramis elongatis simplicibus, cymis fasciculiformibus sessilibus,* 2-10-*floris, juxta totam ramorum longitudinem dispositis. Hügel.*

Macarthuria *australis. Hügel, Enum. Pl. Nov. Holl. p.* 11.

HAB. Australia, King George's Sound. *(Hügel.)* Swan River. *Drummond.*

Hügel observes that this genus ranks between *Thomasia* and *Seringia ;* differing from the latter in the persistent calyx, and the presence of petals, by the stamens being all fertile and the anthers opening with lateral dehiscence ; and from the former by having the stamens united into a cup, by the styles being distinct at the base, and in the absence of bracteas beneath the calyx ;—while from both of these it is very distinct in habit.

Fig. 1. Flower. *f.* 2. Stamens and pistil. *f.* 3. pistil. *f.* 4. ovary cut through transversely. *f.* 5. Capsule. *f.* 6. Capsule burst open. *f.* 7. seed :—*magnified.*

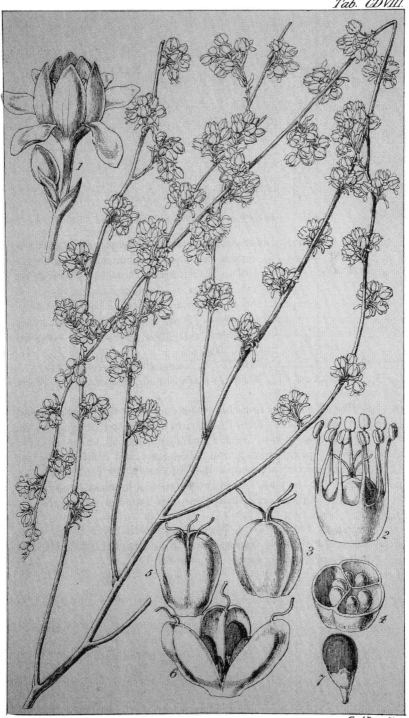

TAB. CDIX.

POLYPODIUM (Dictyopteris) ATTENUATUM. *Br.*

Caudice repente radicibusque ferrugineo-tomentosis, frondibus
  simplicibus aggregatis submembranaceis lanceolatis costatis
  basi in petiolum longe attenuatis reticulatis, areolis oblongis,
  soris ellipticis oblongis uniserialibus anastomosi venularum
  insidentibus.
Polypodium attenuatum. *Br. Fl. Nov. Holl. p.* 147. *Spreng.*
  Syst. Veget. v. 4. *p.* 56.
P. Brownianum. *Spreng. (fide Presl).*
Dictyopteris attenuata. *Presl. Tent. Pterid. p.* 194. *Hook. Gen.*
  Fil. tab. LXXI. B.
HAB. New Holland. *Brown.* New Zealand. *All. Cunningham.*
  Wm. Colenso, Esq.

The nature of the venation is of the highest importance in
the study of the ferns; sometimes for discriminating species,
and not unfrequently, especially when combined with difference
in habit, for distinguishing genera. The simply reticulated
venation in this and some allied species, has induced Presl to
constitute the genus *Dictyopteris.* In the present instance the
sori are dense and prominent, the stalks of the sporangia very
long, and they are mixed with articulated filaments or abortive
sporangia.

Fig. 1. Portion of the fertile frond. *f.* 2. Sporangia and
articulated filaments; *magnified.*

Tab. CDIX.

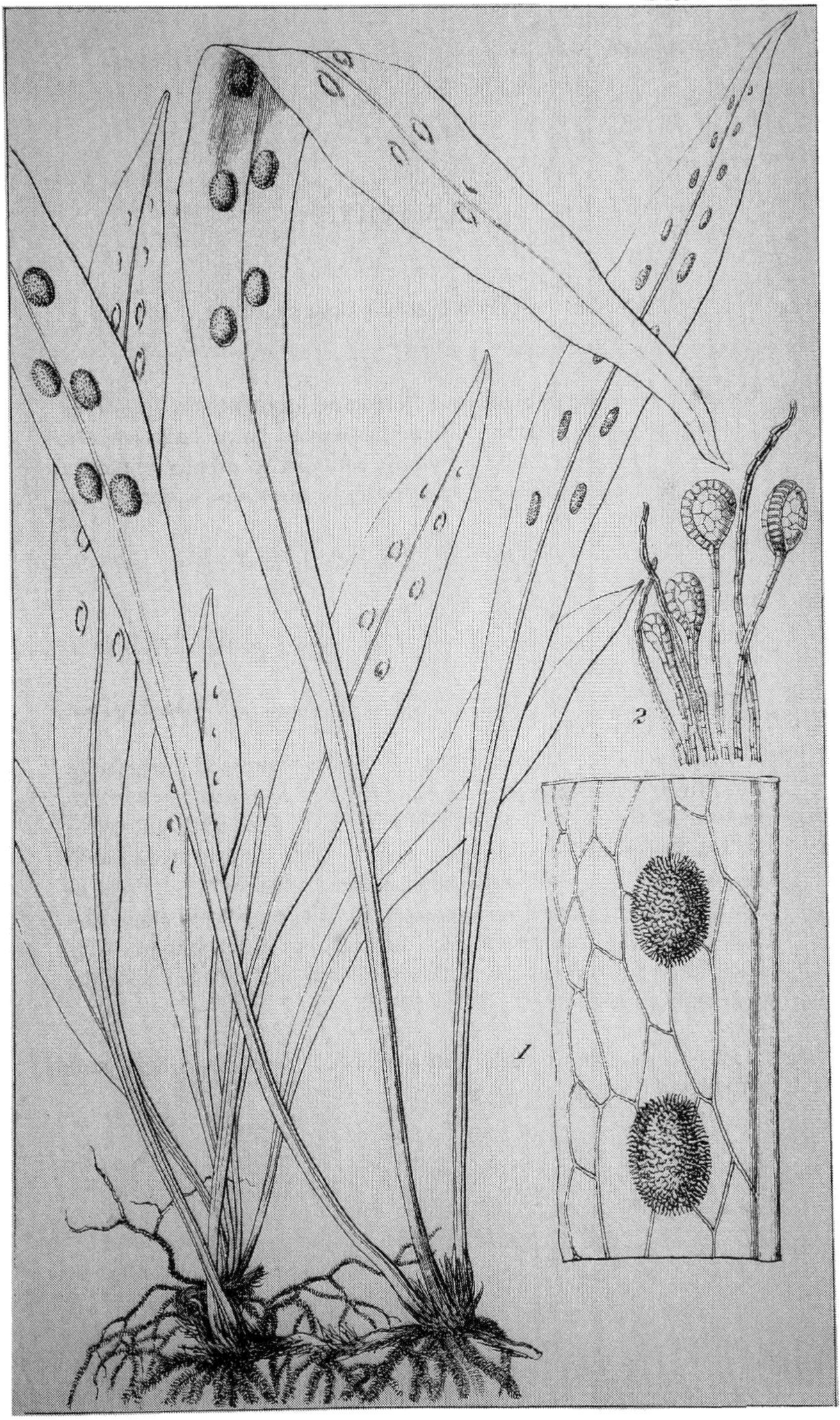

TAB. CDX.

WILSONIA ROTUNDIFOLIA. *n. sp.*

Foliis ovato-rotundatis pilosiusculis, ramis calycibusque sub-cylindraceis dense hirsutis, floribus axillaribus terminalibusque solitariis.

HAB. Australia, Swan River Settlement. *Mr. J. Drummond.*

I am doubtful whether to refer this little plant to Mr. Brown's genus *Wilsonia* or to *Cressa*. It has not the deeply cleft calyx of the latter, nor indeed the urceolate calyx, nor distichous leaves of the former. The true *Wilsonia humilis* of Mr. Brown is figured in our Icones Plantarum at *vol. 3. tab.* 265. The habit indeed of both these plants is extremely similar to that of *Frankenia.* In our plant the cells of the ovary are generally 2-seeded; but the seeds are abortive.

Fig. 1. Flower from the axil of a leaf. *f.* 2. Corolla laid open. *f.* 3. Ovary cut through vertically. *f.* 4. The same cut through transversely. *f.* 5. Anther. *f.* 6. Leaf:—all *magnified.*

Tab. CIX.

TAB. CDXI.

Tropæolum cirrhipes. *n. sp.*

Foliis deltoideis obtusangulis sublonge petiolatis peltatis, pedunculis longissimis filiformibus volubilibus, calycis limbo
erecto in calcar longum subulato-cylindraceum obtusum
attenuato, petalis staminibus styloque inclusis.

Hab. Chacapoyas, Andes of Peru. *Mr. Mathews.* (*n.* 3177.)

I have seen only one specimen of this most remarkable plant,
which in the form of the leaf, and in the extraordinary length
and slenderness of the petiole, is quite unlike any hitherto
described species of the genus. The leaves too have a varie-
gated appearance in the dried state, exhibiting whitish lines, in
which the principal veins run. The calyx and short petals are
yellow-green, the long spur orange-red.

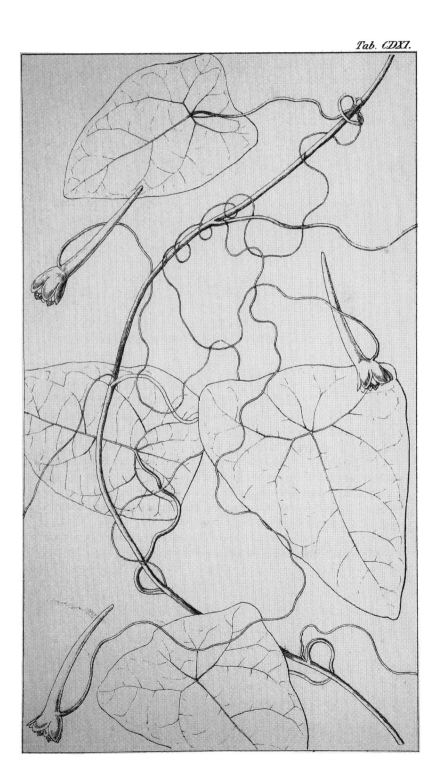

TAB. CDXII.

MACROSTIGMA. *nov. gen.*

GEN. CHAR. Monoca *v.* Polygama. *Calyx* unibracteatus, monophyllus, subturbinatus, persistens, quinquelobus, lobis obtusis margine ciliatis. *Corolla* o.-HERMAPHR. *Stamina* 10 exserta, hypogyna. *Filamenta* libera, glabra. *Antheræ* filamentis longiores, oblongæ, acutæ, minute glandulosæ, biloculares, lateraliter et longitudinaliter dehiscentes. *Germen* obovatum, uniloculare, biovulatum, ovulis ad basin loculi erectis. *Stylus* basilaris, sursum curvatus, dein deflexus, germine subtriplo longior. *Stigma* maximum, peltatum, granulosum, germinis fere magnitudine.—FŒM. *Stamina* 10 abortiva, ad filamenta elongata flexuosa antheris destituta redacta. *Pistillum* ut in hermaphrodita.—Frutex *erectus ramosus, ramis virgatis; foliis sparsis linearibus rigidis obtusis, basi utrinque stipula minuta brevi-subulata suffultis.* Flores *axillares in foliorum axillis superiorum, v. si mavis, racemosi, racemis foliosis.* Macrostigma *australe.*

HAB. Swan River Colony, Australia. *Mr. James Drummond.*

On the Natural Order to which this may be referred I will not venture to offer a conjecture, but content myself with representing such an analysis of this singular plant as my specimen will allow.

Fig. 1. Hermaphrodite flower. *f.* 2. The same, the calyx laid open. *f.* 3. Pistil. *f.* 4. Ovary laid open. *f.* 5, 6. Stamens. *f.* 7. Female flower :—*magnified.*

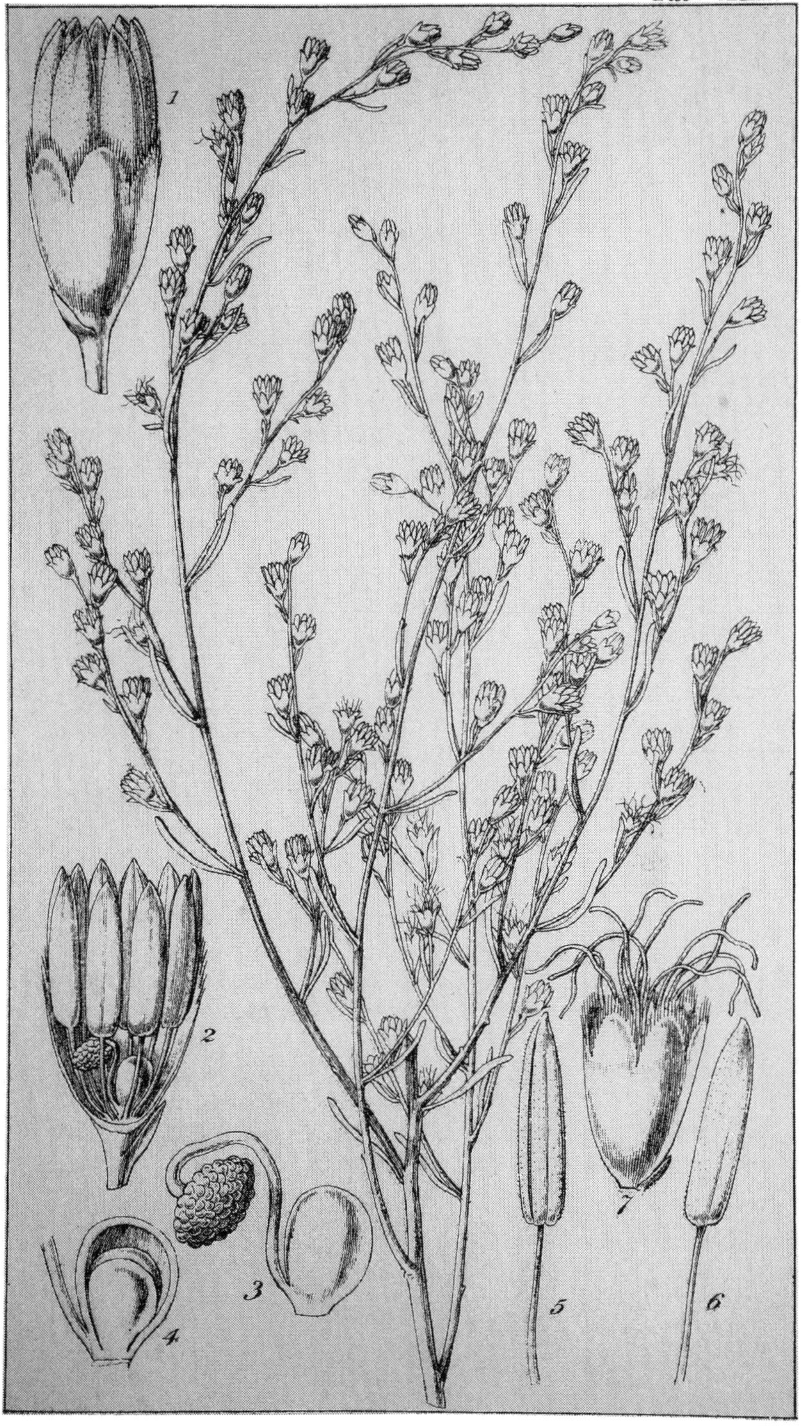
1
2
3
4
5
6
7

TAB. CDXIII.

CROSSOLEPIS? PUSILLA. *Hügel.*

Erecta, glomerulis oblongis basi attenuatis, capitulis bifloris. *Hügel, Enum. Pl. Nov. Holl. p.* 61.

HAB. Swan River Colony, Australia, *(Hügel), Mr. J. Drummond.*

A small erect annual plant, branching from the base : the stems red, clothed with deciduous down. Leaves alternate, linear, very obtuse. Capitula terminal, collected together so as to form a dense cylindrical spike of a glossy, straw-colour, attenuated at the base. Each capitulum consists of a two-flowered involucre of three scales, of a very delicate, membranaceous reticulated texture : the outer one (comparatively) large, almost orbicular, concave, denticulate at the margin : the two inner small, boat-shaped, compressed, so as to present a flattened keel, fringed at the margin above. Within the fold of each of these small scales is a very minute tubular floret. Ovary obovate, tuberculate. Corolla funnel-shaped, widening upwards, 3-lobed. Anthers and style altogether included.

Fig. 1. Spike or glomerule of capitula. *f.* 2. Inner view of a capitulum. *f.* 3. Outer view of do. *f.* 4, 5. The two inner scales of the capitulum with the flowers. *f.* 6. floret. *f.* 7. Upper part of the corolla laid open. *f.* 8. Branches of the style : *magnified.*

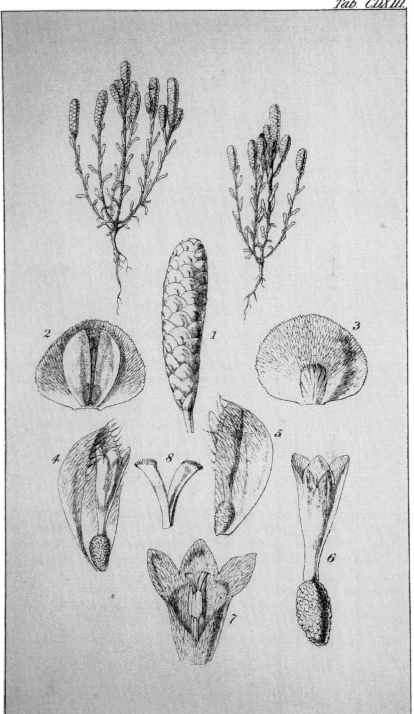

TAB. CDXIV.

LACHNOSTACHYS. *n. gen.*

Flores hermaphroditi bracteati. *Perianthium* longissime densissimeque lanatum, monophyllum, 6-lobatum, scariosum. *Stamina* hypogyna 6-8, perianthii lobis opposita. *Filamenta* filiformia, in tubum basi vel usque ad medium connata, tubo intus villosissimo. *Antheræ* biloculares. *Ovarium* subglobosum. *Stylus* filiformis. *Stigma* obtusum.—Frutices *oræ occidentalis Novæ Hollandiæ; ubique tomentosæ.* Folia *opposita rigida.* Spicæ *terminales et axillares, bracteatæ, cylindraceæ.* Flores *lana longissima ramosa intertexta tecti.*

Lachnostachys *albicans;* foliis lanceolato-ellipticis imbricatis ramisque albo-tomentosis, bracteis flore brevioribus, perianthio 6-lobo, staminibus exsertis, tubo filamentis subæque longo, ovario superne densissime piloso.

HAB. Swan River Colony, Australia. *Mr. James Drummond.*

Two very remarkable plants in Mr. Drummond's Swan River collection are those figured in the present and succeeding plate, belonging to the Order *Amaranthaceæ;* but so different from any genus known to me, especially in habit, that although my specimens are destitute of fruit, and although, on account of the singularly dense and intricate nature of the wool which covers the flowers, it is exceedingly difficult to investigate the exact structure of the minute flowers concealed within the woolly covering, I have ventured to constitute of them a new genus.

Fig. 1. Flowers. 2. Perianth laid open, the stamens being removed. *f.* 3. Stamens. *f.* 4. Pistil. *f.* 5, 6. Anthers. *f.* 7. Small portion of wool from the perianth : all more or less *magnified.*

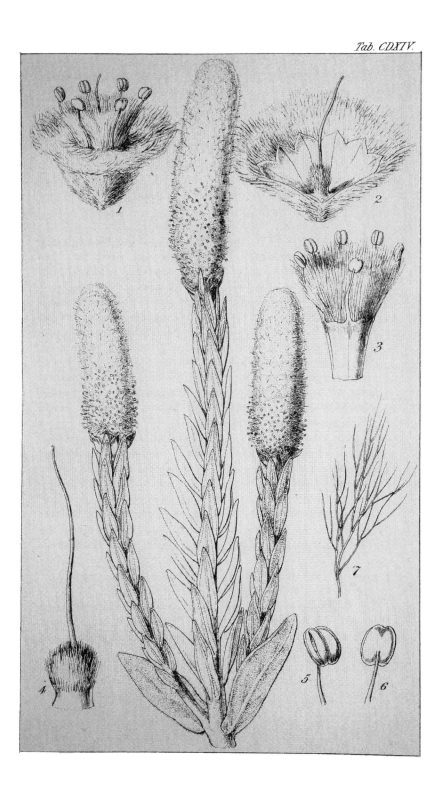

TAB. CDXV.

LACHNOSTACHYS FERRUGINEA. *Hook.*

Foliis lato-ellipticis remotis ramisque dense ferrugineo-tomentosis, bracteis flores superantibus, perianthio 8-lobo, staminibus inclusis, tubo filamentis breviore, antheris dorso tuber-
culato, ovario granulato.

HAB. Swan River Colony, Australia, *Mr. J. Drummond.*

It is possible that when this and the preceding plant *(L. albi-*
cans) are better known as to the structure of their fructification,
the present may be found to constitute a different but closely allied
genus. The bracteas are very large and of a ferruginous brown
colour, contrasting singularly with the dense white balls of
wool which cover the flowers within the bracteas; the perianth
has 10 lobes or segments; the stamens are 8; the tube short;
and at the back of each anther is a large granulated excrescence.

Fig. 1. Flower. *f.* 2. Perianth and staminal tube laid open.
f. 3. Stamens and pistil. *f.* 4, 5. Anthers. *f.* 6. Small portion
of wool from the perianth. *f.* 7. Inner; and *f.* 8, an outer
view of a bractea: all more or less *magnified.*

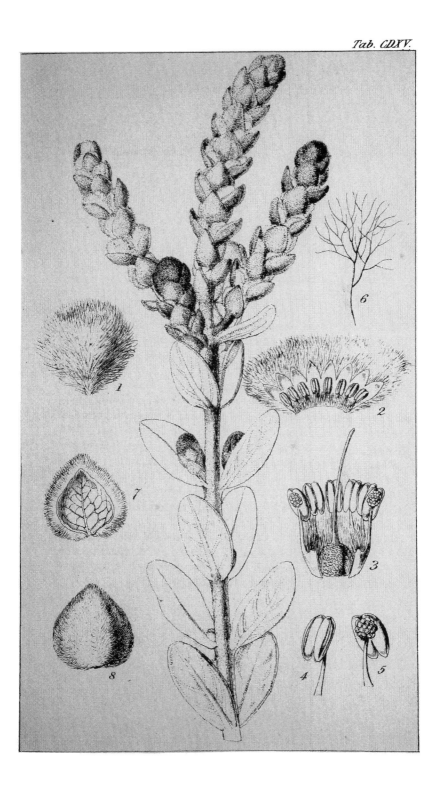

TAB. CDXVI.

Triglochin? calcaratum. *n. sp.*

Triandrum, sepalis 3 ext. calcaratis, carpellis semiunitis, 3 ext. fertilibus apice reflexis, foliis linearibus flaccidis scapo brevioribus, floribus laxe spicatis.

Hab. Swan River Colony, Australia. *Mr. James Drummond.*

Radix fibrosa. Folia 3-5 uncias longa, linearia, flaccida, basi dilatata, membranacea. Scapi 5-6 uncias longi, graciles, flaccidi. Flores laxe spicati. Sepala 6 erecta, quorum 3 exteriora majora, lato-ovata, acuta, antherifera, basi calcarata ; 3 interiora ovata, ecalcarata. Ovaria 6 ovato-acuminata, primum erecta, subæqualia, inferne coadunata : tria exteriora fertilia, demum (statu fructificante) superne reflexa, stigmate infra apicem notata; tria interiora abortiva semper erecta. Ovulum solitarium, oblongum, erectum.

Fig. 1. Flowers. *f.* 2. Outer sepal, with its anther. *f.* 3. Front view of an anther. *f.* 4. Flower from which the 3 outer sepals are removed. *f.* 5. The pistils. *f.* 6. Inner sepal. *f.* 7. Immature fruit. *f.* 8. One of the outer carpels. *f.* 9. The same, the cell laid open. *f.* 10. Immature seed. *f.* 11. The 3 inner or abortive carpels :—all more or less *magnified.*

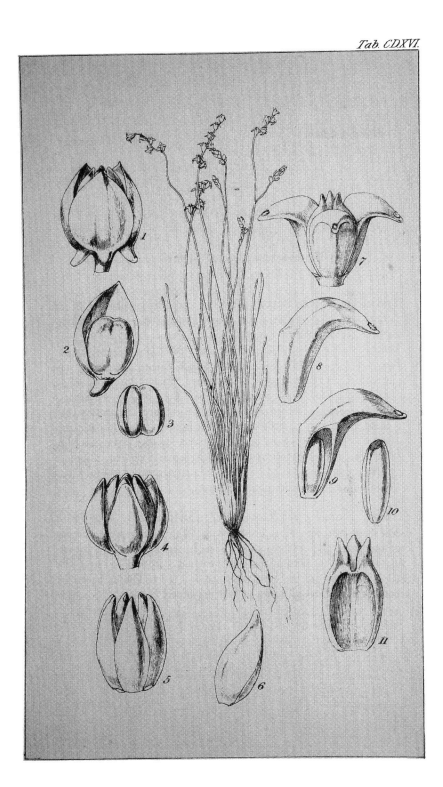

TAB. CDXVII.

LAWRENCIA GLOMERATA. *n. sp.*

Ramosissima, foliis spathulatis petiolatis superioribus sessilibus, stipulis ovatis acutis adnatis, floribus 2-3 glomeratis axillaribus, calyce plicato, carpellis reticulatim venosis.

HAB. Swan River Colony, Australia. *Mr. James Drummond.*

At Tab. CCLXI. of vol. 3 of this work, I established the genus *Lawrencia,* upon a very remarkable plant found on the northern coast of Van Diemen's Land and the opposite southern extremity of Australia, *Lawrencia spicata.* The present Swan River plant is undoubtedly a second species of the same genus.

The lower part of the stem seems to be woody, the rest herbaceous, much branched. Leaves with persistent adnate stipules, which are large and very distinct in the upper floral leaves. The flowers are axillary, glomerate; the calyx singularly plicate in the sinuses, the lobes very acute, erect. Petals acute, combined by their claws with the base of the staminal tube. The styles are 5. Carpels 5, adnate, the sides strongly reticulated. Different as the two species of *Lawrencia* are in habit from *Sida,* the structure of the flowers and fruit is nearer to that genus than I had imagined.

Fig. 1. Flower and bracteas. *f.* 2. Corolla. *f.* 3. Stamens. *f.* 4. Immature carpels. *f.* 5. Single ripe carpel. *f.* 6. The same laid open. *f.* 7. Seed. *f.* 9. leaf:—all more or less *magnified.*

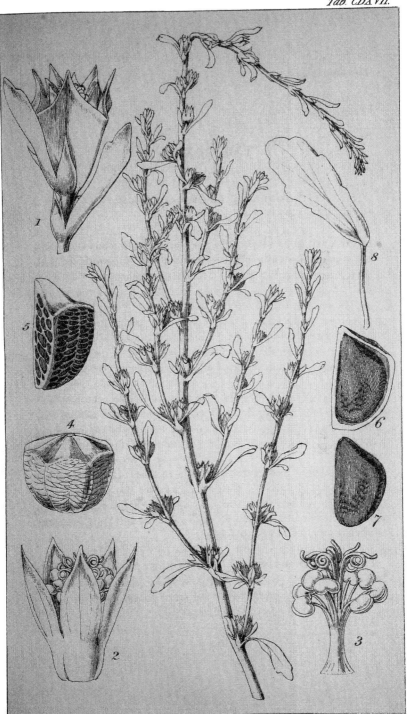

TAB. CDXVIII.

OXALIS CATARACTÆ. *All. Cunn.*

Cæspitosa ramosa decumbens, foliis longe petiolatis, foliolis sessilibus obcordato-lobatis lobis subdivergentibus, adultis utrinque caulibusque glabris venosis, subtus albido-glaucis, marginibus revolutis subintegris, petiolis (uncialibus) membranaceis basi dilatatis scariosis semivaginantibus, pedunculis elongatis unifloris petiolo longioribus pilis albidis conspersis, calycibus pilosis corolla fere triplo brevioribus. *All. Cunn.*
Oxalis Cataractæ. *All. Cunn. Bot. of N. Zeal. in Ann. of Nat. Hist. v.* 3, *p.* 315.

HAB. Northern Island of N. Zealand, on rocks beneath the great fall of the Kerri-Kerri river. *A. and R. Cunningham, W. Colenso, Esq.*

My specimens of this pretty little *Wood-sorrel* do not indeed exhibit the branching nature of the decumbent stem, but that it is the *O Cataractæ* of Mr. A. Cunningham I cannot doubt, since it was sent me under that name, by Mr. Colenso, who gathered it in company with that lamented botanist. It is remarkable for the very large membranaceous stipules which form conspicuous sheaths around the slender stem.

d

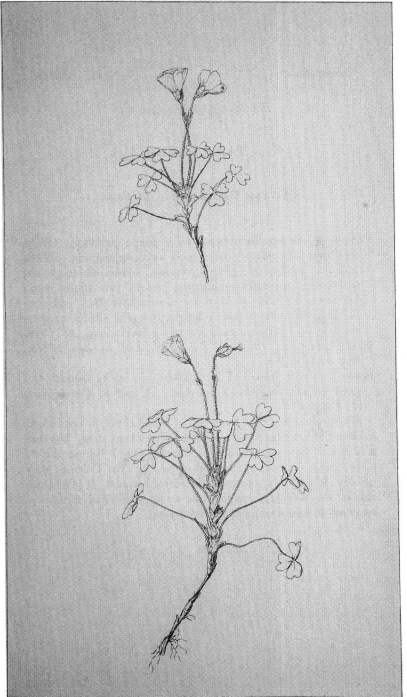

TABS. CDXIX, CDXX.

VITEX LITTORALIS. *A. Cunn.*

Foliis ternatis quinatisve, foliolis ellipticis obtusis cum acumine petiolatis glabris, paniculis brevibus racemosis axillaribus terminalibusve, ramis dichotomis, calyce campanulato subdentato, staminibus exsertis, corolla extus tomentosa. *A. Cunn.*
Vitex littoralis. *All. Cunn. Bot. of N. Zeal. in Ann. of N. Hist. v.* 1, *p.* 461.
Ephialis pentaphylla. *Banks et Sol. Mss. et Ic. ined. in Biblioth. Banks. (A. C.)*
HAB. Rocky shores of the Bay of Islands, N. Zealand, growing frequently within the range of salt water. *All. Cunningham, Mr. Colenso.*

This is described as a tree of very irregular growth, and which, from the hardness and durability of its wood, has been denominated the *New Zealand Oak,* and indeed it seems to answer all the purposes of that prince of trees. The wood is of a dark brown colour, close in the grain, and takes a good polish. It splits freely, works well, and derives no injury from exposure to the damp; twenty years' experience having proved that, in that period, it will not rot, though in a wet soil and underground. For ship-building it is, like the Teak (which belongs to the same natural order), a most valuable wood. It grows from 15 to 30 feet high without a branch, and varying from 12 to 20 feet in circumference.

Fig. 1. Flower. *f.* 2. Stamen. *f.* 3. Pistil: *magnified.*

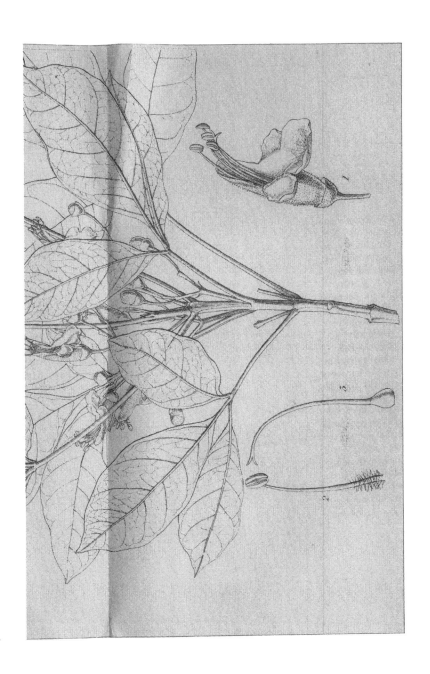

TAB. CDXXI.

FUCHSIA PROCUMBENS. *R. Cunn.*

Apetala, caule procumbente, foliis parvis longe petiolatis alternis cordato-rotundatis denticulatis, pedunculis solitariis axillaribus unifloris petiolo floreque brevioribus, calycis lobis oblongis reflexis, tubo superne dilatato, staminibus exsertis, stylo stamina superante, stigmate capitato.

Fuchsia procumbens. *R. Cunn. mst. All. Cunn. Bot. of N. Zeal. in Ann. of N. Hist. v.* 3, *p.* 31.

HAB. Northern Island, N. Zealand, around the village of Matauri on the East Coast, opposite the Cavallos Isles, inhabiting the sands immediately above the range of the tide, where it was found in flower in March, 1834, by *Richard Cunningham.* Found also by *W. Colenso, Esq.* to whom I am indebted for the specimen here figured.

This is very different from the only other species of the genus yet known to inhabit N. Zealand, and from every other described one. We have seen a living plant of it in the possession of the Rev. Mr. Williams of Hendon.

Fig. 1. Flower: *magnified.*

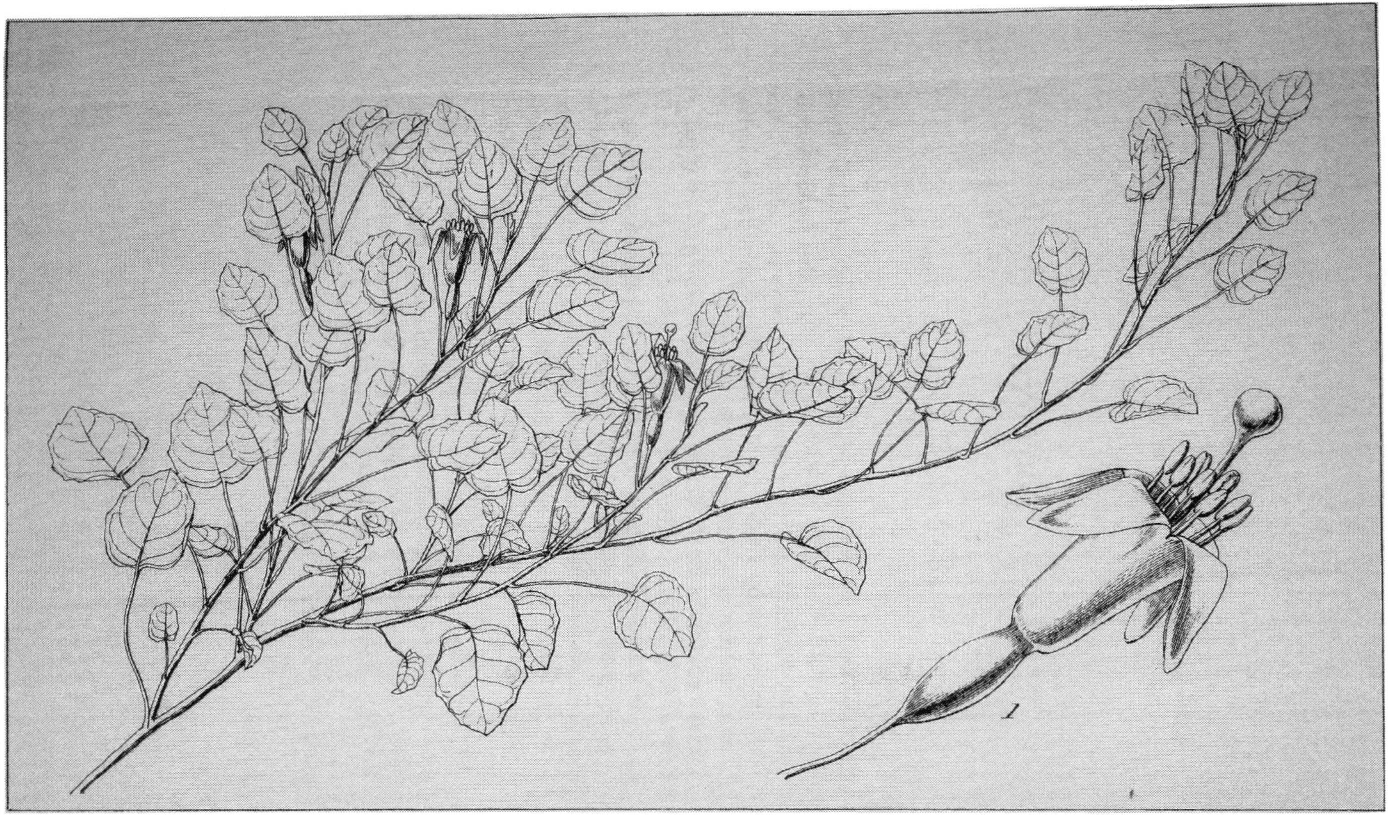
1

TAB. CDXXII.

Pteris (Allosorus) rotundifolia. *Forst.*

Frondibus pinnatis, pinnis alternis obtusissimis cum mucrone glabris obsolete nervosis, superioribus ovato-ellipticis basi truncatis, inferioribus rotundatis basi cuneatis, soris latis continuis demum nudis, stipite basi scabro reliquo rachique rufo-hispidis paleaceisque.

Pteris rotundifolia. *Forst. Prodr. n.* 420. *Willd. Sp. Pl. v.* 5, 4. *p.* 563. *Sw. Syn. Fil. p.* 102 *et* 297. *Rich. Fl. Nov. Zeal. p.* 78. *All. Cunn. Bot. N. Zeal. in Hook. Comp. Bot. Mag. v.* 2, *p.* 355.

HAB. New Zealand, Middle Island, *Forster.* Dry forests on the banks of the Kaua-Kaua and Wycady rivers, Bay of Islands. *All. Cunningham, W. Colenso, Esq.* Astrolabe Harbour. *D'Urville.*

This beautiful plant appears to grow in tufts; the fronds, including the stipes, one and a half and two feet high. Stipes and rachis perfectly terete, red-brown, glossy; the base of the former is rough, scarcely hispid, the rest clothed with spreading ferruginous rigid hairs and scales. The pinnæ are about ¾ of an inch long, of a pale very opaque green, paler still below, and there having, generally, a line of paleaceous setæ; the rest quite glabrous and exhibiting no trace of nerves; above, in the dry state, the nerves are very indistinctly seen, pinnated on the costa and dichotomous; and it is on the branches within the margin that the sori form a continued broad line, at first covered with the marginal involucre, afterwards the involucre spreading open and exposing the sori.

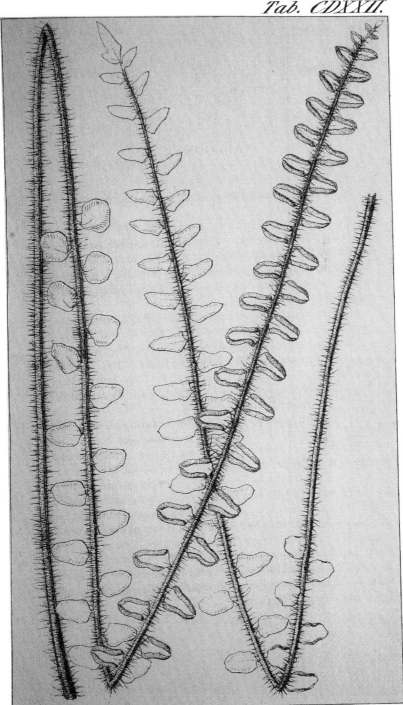

TAB. CDXXIII.

Asplenium bulbiferum. *Forst.*

Frondibus lato-lanceolatis bipinnatis, pinnis alternis lanceolatis glabris, pinnulis ovato-oblongis obtusis inciso-pinnatifidis basi attenuatis in rachi lata decurrentibus, axillis superne proliferis, laciniis integris v. bidentatis, soris in singula lacinia (pinnis inferioribus exceptis) medium versus, stipite rachique alata inferne squamulosis.

Asplenium bulbiferum. *Forst. Prodr. n.* 433. *Willd. Sp. Pl. v.* 5, *p.* 345. *Sw. Syn. Fil. p.* 84, 278. *Schkuhr, Fil. v.* 79. *Spreng. Syst. Veget. v.* 4, *p.* 89. *Rich. Fl. Nov. Zel. p.* 75. (*excl. syn.* A. laxi, *Br.) All. Cunn. Bot. N. Zeal. in Hook. Comp. Bot. Mag. v.* 2, *p.* 364.

HAB. New Zealand. *Forster.* Middle Island, Astrolabe Harbour. *D'Urville.* Northern Island; in humid woods, on the banks of the Kaua-Kaua, Bay of Islands. *All. Cunningham, Wm. Colenso, Esq.*

Our specimens are about 2 feet long. Several of the superior pinnæ, especially in the axils of the segments, bear little bulbs which exhibit themselves on the upper surface and produce young fronds while still attached to the parent.

Fig. 1. Fertile pinnule :—*magnified.*

TAB. CDXXIV.

COROKIA. *A. Cunn.*

GEN. CHAR. *Flores hermaphroditi* (dioici, *A. C.*) *Calycis tubus* elongato-turbinatus, ovario adhærens; limbo 5-fido, persistente, per æstivationem valvato. *Petala* 5, lanceolata, decidua, lobis calycis alterna, intus basi squamula fimbriata instructa. *Stamina* 5, petalis alterna, iis breviora: *Filamenta* basi dilatata: *Antheræ* lineari-oblongæ, intus rimis duabus longitudinalibus dehiscentes. *Glandulæ epigynæ* 5, laciniis calycinis oppositæ: *Ovarium* inferum, biloculare, loculis 1-ovulatis pendulis. *Stylus* staminibus brevior. *Stigma* incrassatum, bifidum. *Drupa* 2-locularis, dipyrena.—Frutex 10-*pedalis, ramulis foliisque subtus albo-tomentosis.* Folia *alterna, coriacea, lanceolata, breviter petiolata, supra glabra, nitida, punctulata, penninervia, reticulata.* Flores *parvi, subpaniculati, bracteati; paniculis brevibus, axillaribus terminalibusque, undique, etiam petalis extus, piloso-canis.*

Corokia buddleoides. *All. Cunn. Bot. N. Zeal. in Ann. of Nat. Hist. v.* 3, *p.* 249.

HAB. New Zealand, Northern Island, margins of woods on the shores of the Bay of Islands, Wangaroa, &c. *A. and R. Cunningham, W. Colenso, Esq.*

The general aspect of this plant is a good deal similar to that of *Buddlea.* Its generic name is derived from that by which it is known to the natives " Korokia-taranga." Mr. Cunningham speaks of it as diœcious. My specimens exhibited stamens in the same flower with the pistil.

Fig. 1. Portion of a leaf, upper surface. *f.* 2. Flower. *f.* 3. Petals and stamens. *f.* 4. Calyx and pistil. *f.* 5. Young fruit. *f.* 6. Ovary cut through. *f.* 7. Young fruit laid open: —*magnified.*

TAB. CDXXV.

Persoonia quinquenervis. *n. sp.*

Ramulis foliisque junioribus alabastrisque sparse pilosulis, foliis spathulato-lanceolatis rigidis mucronatis quinquenerviis sub lente punctulis hyalinis scabriusculis, floribus solitariis erectis, antheris stigmateque obtusis.

Hab. N. Holland, Swan River Colony. *Mr. James Drummond.*

With the exception of the young shoots and the flower-buds, which are slightly hairy, the rest of the plant is quite glabrous. The flowers are axillary. Peduncles solitary, single-flowered. Sepals lanceolate, acuminate, coriaceous. Anthers and style glabrous.

Fig. 1. Leaf:—*slightly magnified.*

1

TAB. CDXXVI.

Persoonia Laureola. *Lindl.*

Undique glaberrima, foliis late ovalibus basi angustatis obtuse mucronatis submembranaceis penninerviis, floribus axillaribus erectis, perianthiis acuminatis, antheris obtusis, stigmate dilatato.

Persoonia Laureola. *Lindl. Sw. Riv. Bot. p.* xxxv.

Hab. Swan River Colony, New Holland. *Mr. Jas. Drummond.*

Allied to *P. salicina*, (Pers. and Brown), but with much broader and thinner, not inæquilateral, leaves. Dr. Lindley describes the apex of the sepals as being minutely pubescent, which is not the case in our specimens.

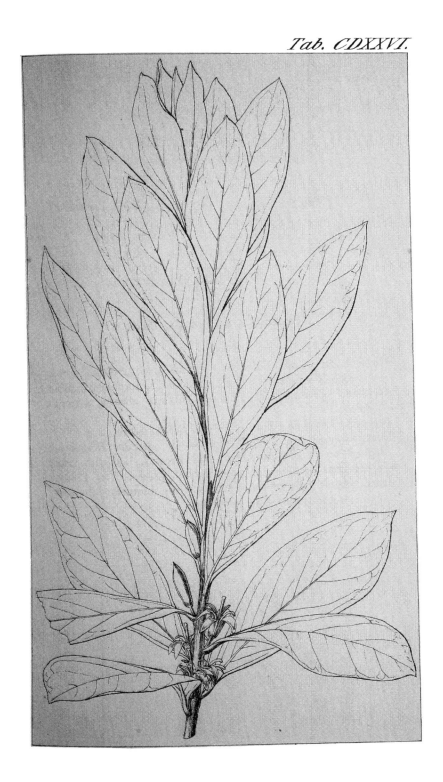

TABS. CDXXVII. CDXXVIII.

LOMARIA PROCERA. (*Spreng.*) *var. β.*

Frondibus pinnatis oblongo-ellipticis, pinnis sterilibus lanceo-lato-ensiformibus acuminatis serratis basi subcordatis, fertilibus (ejusdem v. diversæ frondis) linearibus costa subtus paleacea, indusiis subintramarginalibus. *Br.*

Lomaria procera. *Spreng. Syst. Veget. v.* 4, *p.* 65. *A. Cunn. Bot. of N. Zeal. in Comp. Bot. Mag. v.* 2, *p.* 363. *(excl. syn. Rich.)*

Stegania procera. *Br. Prodr. p.* 153. *(non Rich. Fl. Nov Zel.)*

Blechnum procerum. *Sw. Syn. Fil. p.* 115. *Labill. Nov. Holl.* 2, *p.* 97, *t.* 247. *Willd. Sp. Pl. v.* 5, *p.* 415.

Asplenium procerum. *Bernh. Act. Erf.* 1802, *p.* 4, *f.* 1.

Osmunda procera. *Forst. Prod. n.* 414.

β; pinnis sterilibus valde acuminatis, fertilibus omnibus ad basin soriferis. (TAB. NOSTR. CDXXVII. CDXXVIII.)

HAB. New Holland and Van Diemen's Land. *Brown.* New Zealand. *Forster.* Bay of Islands, Wangaroa, &c. *A. and R. Cunningham, W. Colenso, Esq.*

This appears liable to considerable variation, both in the sterile and fertile pinna. In Labillardière's plant, the former are very obtuse. In a var. mentioned by Mr. A. Cunningham, the base of the fertile pinnæ is much dilated and sterile, similar to what is figured as *Steg. procera* in Rich. *Fl. Nov. Zel.* t. 13. but the sterile frond being there pinnatified, not pinnate, proves that that portion of the plant, at least, cannot be the same as ours.

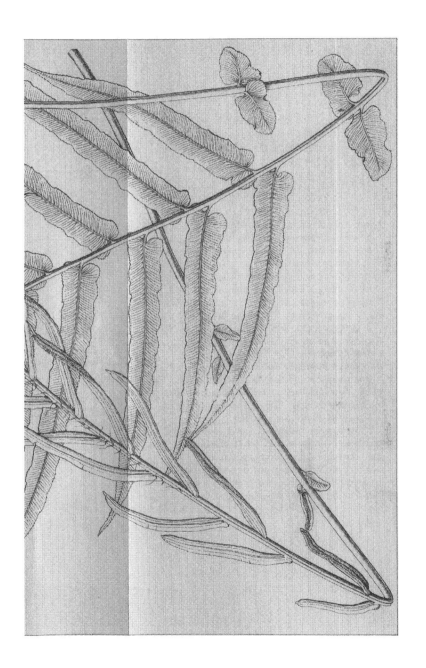

TAB. CDXXIX.

LOMARIA LANCEOLATA. *Spr.*

Frondibus sterilibus pinnatifidis lanceolatis scaberulis laciniis approximatis oblongis obtusiusculis subfalcatis denticulatis infimis abbreviatis orbiculatis, fertilibus pinnatis, pinnis remotis linearibus longitudine fere fertilium, rachi stipiteque nudis.

Lomaria lanceolata.   *Spr. Syst. Veget. v.* 4, *p.* 62.   *All. Cunn. Bot. N. Zeal. in Hook. Comp. Bot. Mag. v.* 2, *p.* 363.

Stegania lanceolata.   *Br. Prodr. p.* 152.   *A. Rich. Fl. Nov. Zel. p.* 86.   *Endl. Prodr. Norf. p.* 81.

HAB. Van Diemen's Land. *Brown.* Norfolk Island. *(Endlicher).* New Zealand, Bay of Islands, Kerri River and Astrolabe Harbour, Cook's Strait. *A. and R. Cunningham, Wm. Colenso, Esq. D'Urville.*

I possess the same, or a very nearly allied species, gathered by Bertero in Juan Fernandez. It scarcely differs, but in the fertile pinnæ being remarkably decurrent, so that the fertile fronds may almost be called pinnatifid.

Fig. 1. Fertile pinna :—*slightly magnified.*

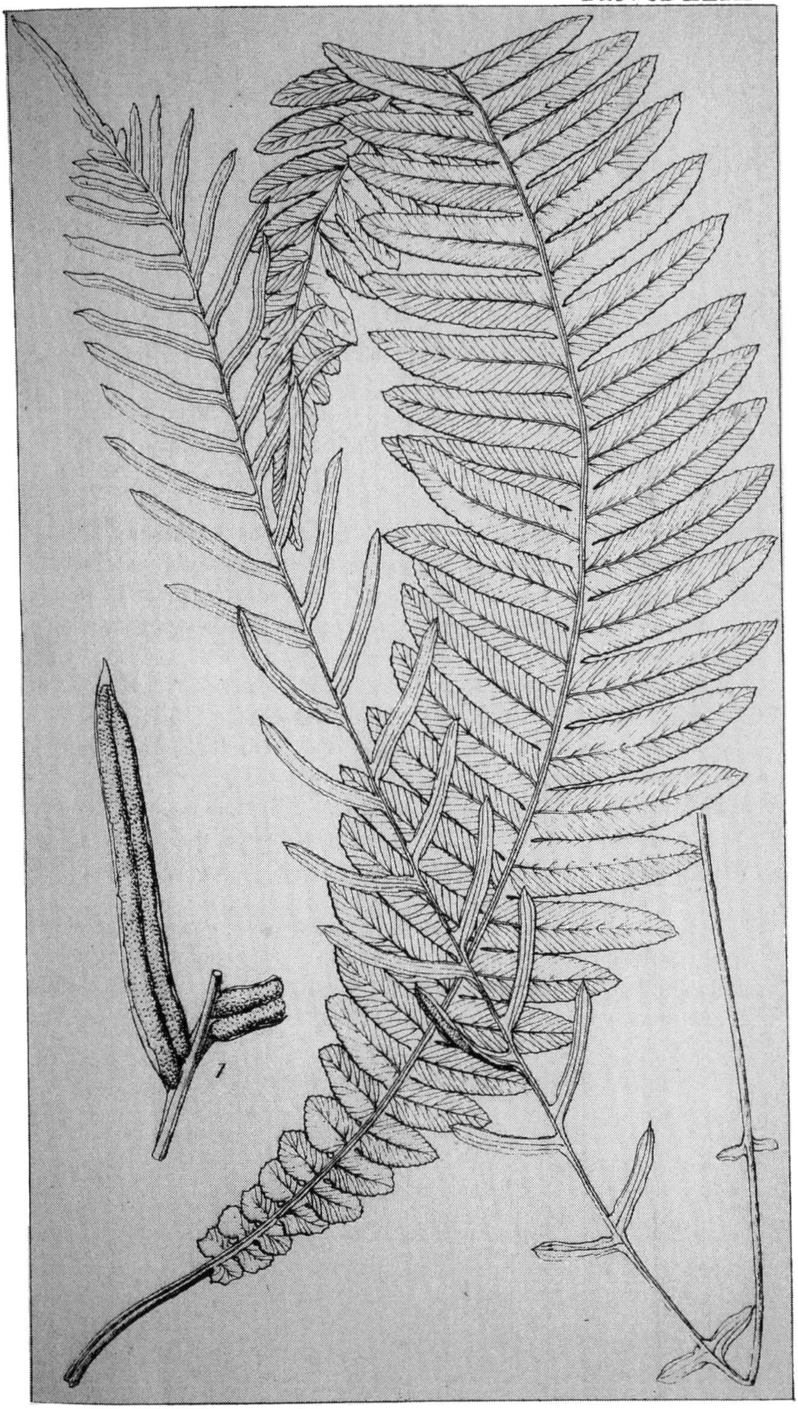
1

TAB. CDXXX.

GENIOSTOMA LIGUSTRIFOLIUM. *A. Cunn.*

Fruticosum, foliis ellipticis ovatisve acuminatis subtus discoloribus, corollæ laciniis reflexis, stigmate depresso-capitato. *A. Cunn.*

Geniostoma ligustrifolium. *A. Cunn. Bot. of N. Zeal. in Ann. Nat. Hist. v. 2, p.* 47.

Geniostoma rupestre. *A. Rich. Fl. N. Zeal. p.* 207. *(non Forst).*

Aspilotum lævigatum. *Banks et Sol. Mss. (fide A. Cunn.)*

HAB. New Zealand, Bay of Islands, in dry woods. *Sir Joseph Banks, All. and R. Cunningham, D'Urville, W. Colenso, Esq.*

Frutex, ut videtur, mediocris, valde ramosus, glaber. Rami teretes. Folia opposita, petiolata, stipulata : stipulis oppositis in vaginulam brevem intrapetiolarem unitis. Pedunculi breves, ramosi, axillares, glomerati, pedicellis bibracteatis. Calyx profunde quinquefidus, inferus, laciniis ovatis, acuminatis, patentibus. Corolla rotato-campanulata, 5-fida, laciniis patentibus vel reflexis, ovatis, intus barbatis. Stamina 5, ad faucem corollæ inserta, laciniis alterna. Filamentum brevissimum : Anthera lato-ovata. Ovarium subglobosum, biloculare; placentis centralibus. Ovula numerosa. Stylus brevis. Stigma capitatum, medio depressum, subbifidum.

Fig. 1. Flower. *f.* 2. Calyx and pistil. *f.* 3. Stamen. *f.* 4. Ovary cut through transversely :—*magnified.*

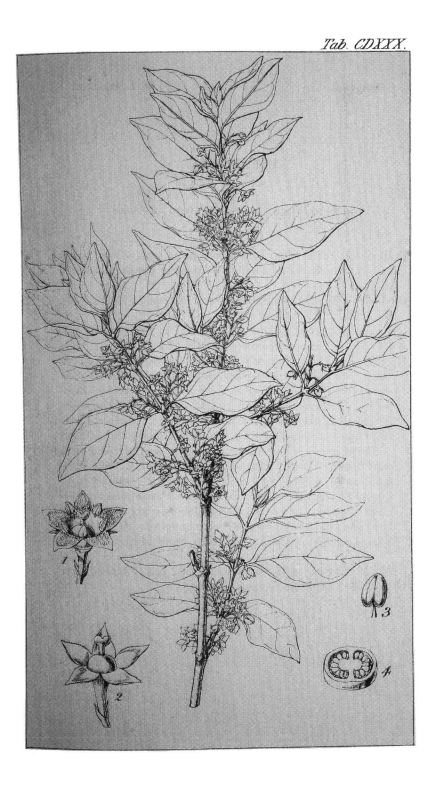

TAB. CDXXXI.

EARINA. *Lindl.*

GEN. CHAR. *Sepala* erecta, æqualia, acuta, membranacea, carinata. *Petala* carnosa, obtusata. *Labellum* carnosum, posticum, cucullatum, trilobum, disco nudo, cum columnâ continuum et subparallelum. *Columna* teres, nana, stigmatis obliqui labio inferiore prominulo. *Clinandrium* proclive. *Anthera* bilocularis. *Pollinia* 4, preparia cohærentia, collateralia. —Herba *caulescens; rhizomate articulato, repente.* Folia *linearia, disticha, vaginantia.* Flores *parvi, paniculati, bracteis cartilagineis, striatis, auriculatis. Lindl.*

Earina mucronata.   *Lindl. in Bot. Reg. sub t.* 1699.
Epidendrum autumnale.   *Forst. Prodr. n.* 319.
Cymbidium autumnale.   *Sw. Nov. Act. Ups. p.* 72.   *Rich. Fl. N. Zel. p.* 169.

HAB. New Zealand, Northern Island, *Sir Jos. Banks.* Moist woods, on rocks and trees, Bay of Islands, Wangaroa, &c. *A. and R. Cunningham, W. Colenso, Esq.* Dusky Bay. *Forster.*

I believe the general structure of the flower, as here represented, is correct ; but the specimens did not allow of so accurate an analysis as I could have wished. Professor Lindley refers the genus to the group of *Malaxideæ.*

Fig. 1. Flower. *f.* 2. Labellum. *f.* 3. Column :—*magnified.*

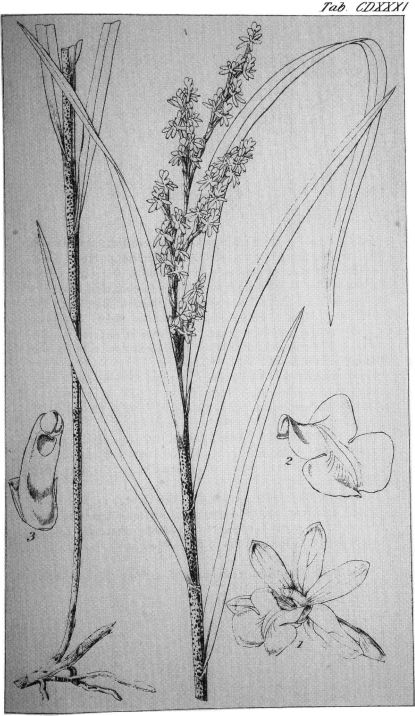

TAB. CDXXXII.

HAKEA CONCHIFOLIA. *n. sp.*

Ramis superne dense pubescentibus hirsutissimisque, foliis reniformi-cordatis repandis spinoso-dentatis reticulatim venosis glaucis, floribus axillaribus fasciculatis.

HAB. New Holland, Swan River Colony. *Mr. Jas. Drummond.* A species evidently nearly allied to *Hakea cucullata, Br. Prod. Suppl.* p. 30, detected by Mr. Baxter, at King George's Sound: but that has the leaves quite destitute of spinous teeth. The fruit I have not seen. The flowers are small, and in the dried specimens at least, concealed by the concave and almost convolute leaves.

Fig. 1. Flowers. *f.* 2. Single flower more expanded. *f.* 3. Pistil with the hypogynous gland. *f.* 4. Apex of a sepal, with the anther :—*magnified.*

TAB. CDXXXIII.

HAKEA PLATYSPERMA. *n. sp.*

Foliis tereti-filiformibus apice mucronato-spinosis, capsulis glo-
boso-compressis ecalcaratis rugosulis, valvis exacte hemisphæ-
ricis intus concavis cribrosis, seminibus orbicularibus latissi-
me alatis hinc lævibus illinc disco muricatis.
HAB. Swan River Colony, New Holland. *Mr. Jas. Drummond.*

The fruit, perhaps, of the *Hakeæ* in general, will be found to
afford excellent characters for distinguishing the species: and
here the capsule is very remarkable and very much resembling
castanets. Each valve is hemisphærical, concave within, and
there having several irregular openings, 2 or 3 lines deep: these
are occupied by the spine-like processes of the back of each seed:
and these seeds are so large as to fill the whole area of the valves.

Fig. 1. Inner view of a seed. *f.* 2. Outer view of do. showing the muricated disk. *f.* 3. Side view of a seed, showing the
smooth *inner*, and the muricated disk of the *outer* surface:—
nat. size.

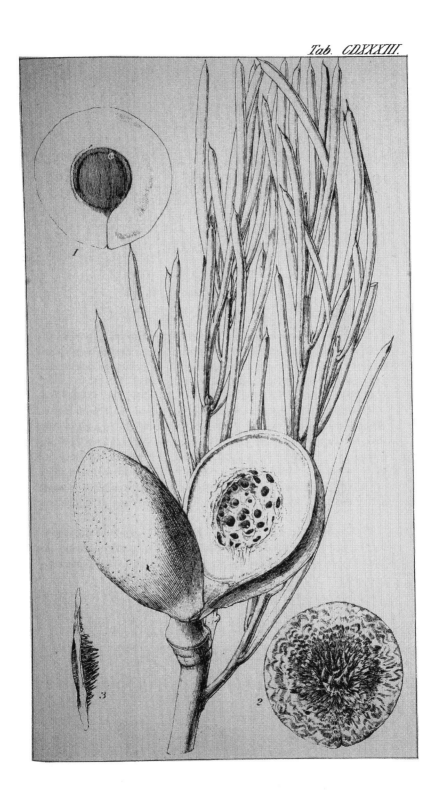

TAB. CDXXXIV.

HAKEA PANDANICARPA. *Br.*

Foliis integerrimis oblongo-linearibus basi attenuatis immerse nervosis aveniis apiculo sphacelato, capsulis gibbosis obovatis tessellatis tuberculis conicis, seminibus undique alatis. *Br.*

Hakea pandanicarpa. *Br. Prodr. Suppl. p. 29.*

HAB. Between Cape Arid and Lucky Bay, South shores of N. Holland. *Mr. Baxter.*

This is very appropriately named by Mr. Brown, from the resemblance of its fruit to that of a *Pandanus* (Screw Pine). I have not seen the flowers, nor does it appear that they were discovered.

TABS. CDXXXV. CDXXXVI.

HAKEA TRICOSTATA. *Br.*

Ramis gemmisque tomentosis, foliis oblongis obtusis mucronatis grosse trinerviis venosis marginatis inferne attenuatis
junioribus sericeis, capsulis erectis ovatis acuminatis ecalcaratis tuberculatis, pedunculo fructifero brevi superne incrassato.
HAB. King George's Sound. *Mr. Baxter.*
I do not find any species in Mr. Brown's Prodromus (including
the Supplement) which accords with this. The leaves are 5-7
inches long, thick and hard. In the axils of the upper ones are
the floriferous gemmæ. Lower down are the ripe capsules,
scarcely an inch long, with a short thickened peduncle, and
beset with small scattered dark-coloured warts.

TAB. CDXXXVII.

HAKEA HETEROPHYLLA. *n. sp.*

Foliis mucronatis tereti-filiformibus compressis hinc sulcatis simplicibus vel bi-trifurcatis, aliis ovali-spathulatis planis, gemmis floriferis terminalibus, capsulis deflexis oblique ovatis compressis tuberculatis in ramis brevibus terminalibus.

HAB. Swan River, New Holland. *Mr. Fraser.*

There are only three species in that division of *Hakea* to which this plant belongs, "*Folia plura filiformia: aliqua plana.*" Two of them are from the south coast of New Holland, but neither agrees precisely with the present, which has three very distinct forms of leaf; 1. tereti-filiform, compressed, with a groove on the upper side; 2. more compressed, and bi-trifurcate or subpinnatifid ; 3. shorter, broadly spathulate and quite entire. The floral gemmæ are on short, patent branches, and the capsules are also terminal on the older and thicker ones.

f

Tab. CDXXXVII.

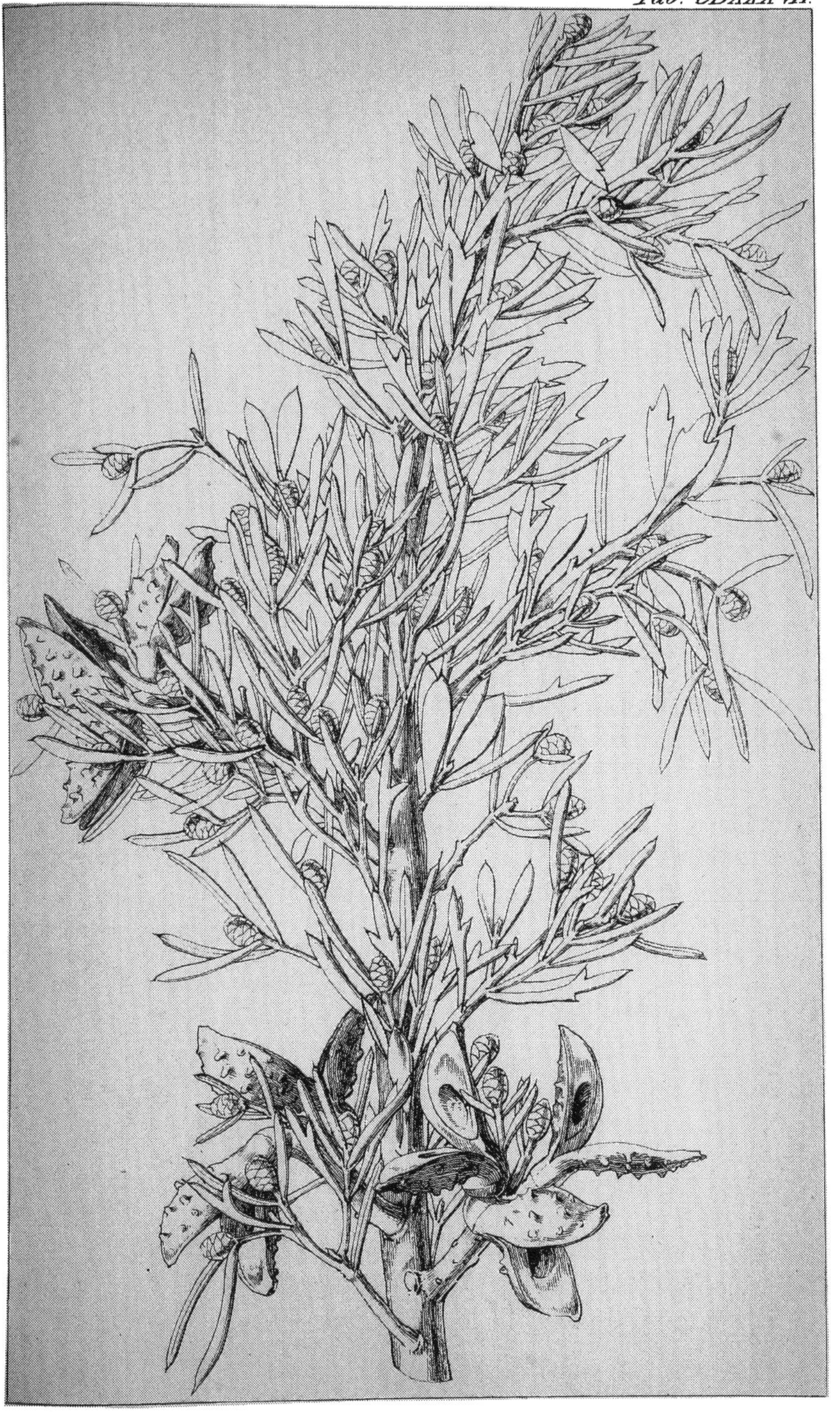

TAB. CDXXXVIII.

ISOPOGON AXILLARIS. *Br.*

Foliis cuneato-lingulatis mucronulatis, capitulis axillaribus
paucifoliis, bracteis involucrantibus ovatis imbricatis, perianthii laminis longitudinaliter barbatis, stigmate fusiformi.
Br.

Isopogon axillaris. *Br. Linn. Trans. v.* 10. *p.* 74. *Prodr. p.* 367.
HAB. South coast of New Holland. *R. Brown, Esq.* King
George's Sound. *Fraser.*

This, in its inflorescence, differs considerably from the
greater number of species of *Isopogon.* Here the flowers are
axillary and lax. Each segment of the perianth, too, has a
beautiful tuft of white silky hairs, and the stigma is fusiform.

Fig. 1. Flower. *f.* 2. Pistil. *f.* 3. Apex of a segment of
the perianth : *magnified.*

TABS. CDXXXIX. CDXL.

HAKEA BAXTERI. *Br.*

Foliis flabellato-cuneatis apice rotundato multidentato lateribus integerrimis, adultis glabris immerse venosis, capsulis ecalcaratis gibbosis. *Br.*

Hakea Baxteri. *Br. Prodr. Suppl. p.* 28.

HAB. New Holland, King George's Sound. *Mr. Baxter.*

Nothing can be more singular than the varied form of the fruit and foliage of the genus *Hakea*, of which numerous species exist on the south and south-western shores of Australia. The present has beautifully fan-shaped leaves, but of a singularly thick and coriaceous character.

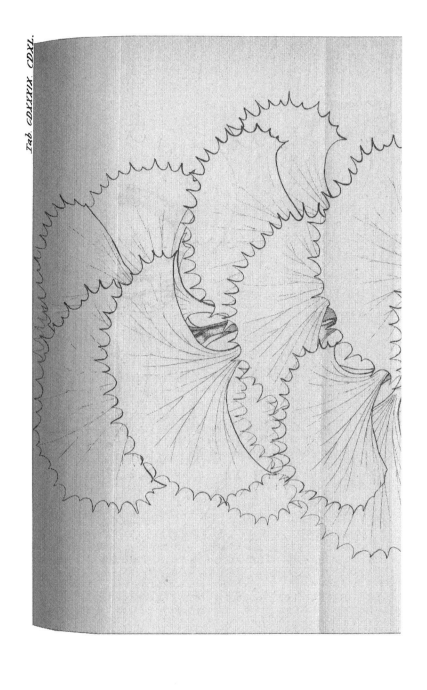

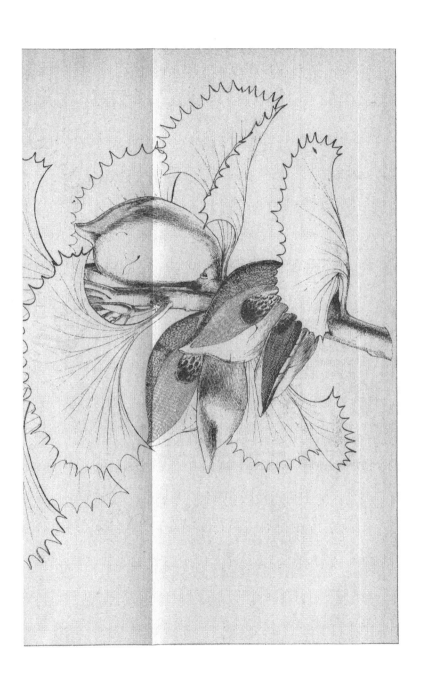

TAB. CDXLI.

HAKEA CUCULLATA. *Br.*

Foliis integris subrepandis cucullatis reniformi-cordatis acutiusculis nervosis reticulato-venosis, capsulis ecalcaratis. *Br.*

Hakea cucullata. *Br. Prodr. Suppl. p.* 30.

HAB. New Holland, King George's Sound. *Mr. Baxter.*

The affinity of *H. conchifolia* to this has been already noticed, under our Tab. 432. The fruit only appears to have been discovered of this species.

TAB. CDXLII.

HAKEA INCRASSATA. *Br.*

Foliis integerrimis anguste lanceolatis obsolete 3 (-5)-nervibus apiculo sphacelato, capsulis refractis obovatis (seu obovato-globosis rima longitudinali) gibbosis lævibus ecalcaratis apiculo adscendenti, (pedunculo ramoque fructifero infra capsulam crassissimis). *Br.*

Hakea incrassata. *Br. Prodr. Suppl. p.* 29.

HAB. New Holland, Swan River Colony. *Mr. Fraser,* (fruit.) *Mr. Jas. Drummond,* (flower.)

My fruiting specimen is from Mr. Fraser, to whom Mr. Brown attributes the discovery of this species. But the most remarkable peculiarity about it is the thickening of the fruit-stalk and of the portion of the branch below it, whence I apprehend Mr. Brown's specific name is derived. What I take to be the same species from Mr. Drummond is in flower. The flowers very small, axillary, clustered, downy.

Allen & Ferguson lithog.

TAB. CDXLIII.

Hakea cristata. *Br.*

Foliis cuneato-obovatis spinoso-dentatis immerse venosis ramulisque glaberrimis, capsulis bicristatis, cristis inciso-dentatis. *Br.*

Hakea cristata. *Br. Prodr. Suppl. p.* 28.

Hab. New Holland, Swan River Colony. *Mr. Fraser. Mr. Jas. Drummond.*

The leaves are glaucous, peculiarly harsh and rigid, the bark dark brown, slightly pruinose in the younger branches.

I possess a flowering specimen from the Swan River, with leaves almost twice the size of this, and much broader; the spines more distant, and the bark much paler and redder. The flowers are very small, arising from the axils of deciduous scales (of which the gemmæ are seen in our figure), thus forming a short raceme, of which the axis, or peduncle, is clothed with silky wool.

Tab. CDXLIII

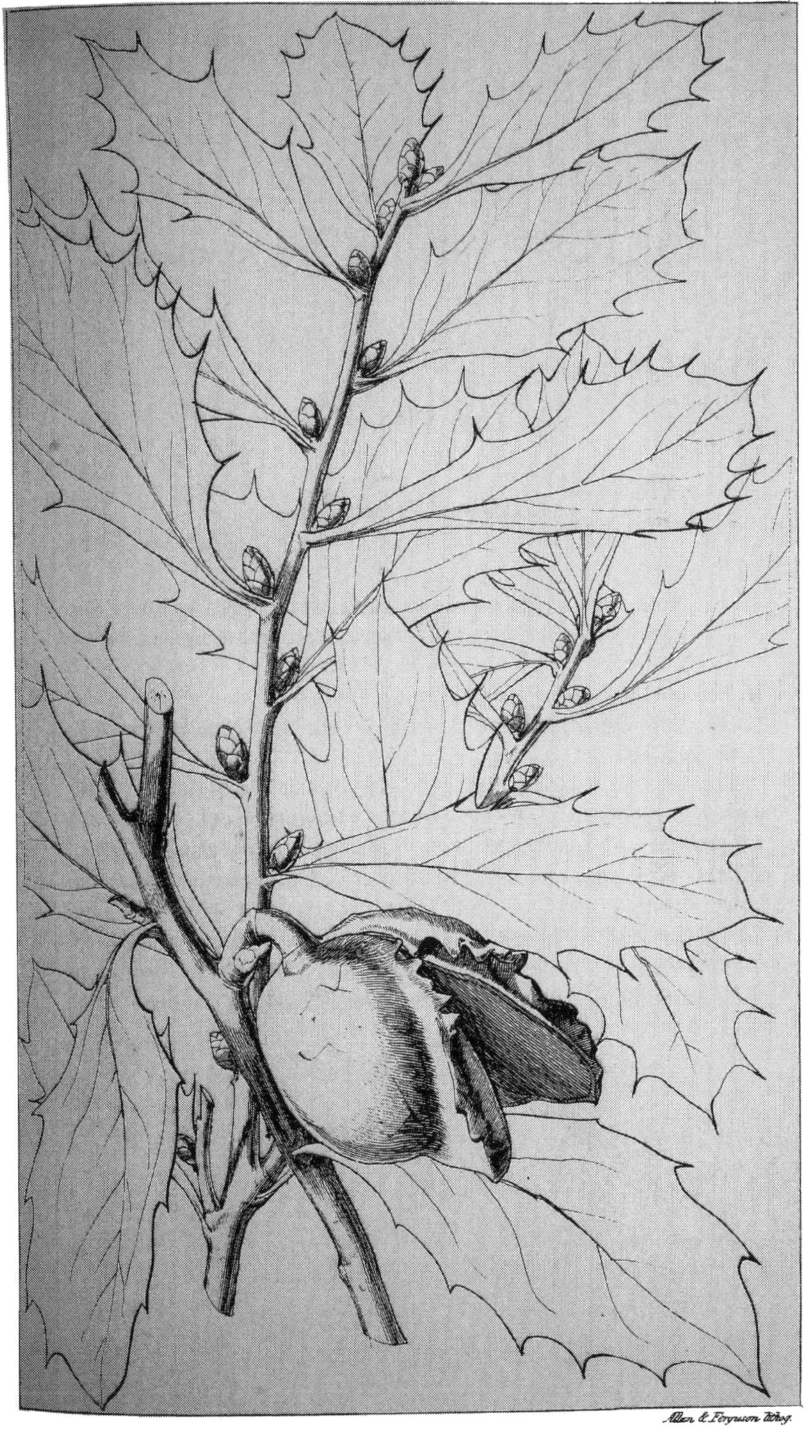

Allen & Ferguson lithog.

TAB. CDXLIV.

HAKEA STENOCARPA. *Br.*

Foliis integerrimis linearibus apiculo acuto sphacelato marginatis uninervibus, venis obsoletis, capsulis lineari-subulatis
falcatis ecalcaratis. *Br.*

Hakea stenocarpa. *Br. Prodr. Suppl. p. 29.*

HAB. New Holland, Swan River Colony. *Mr. Fraser.*

This is remarkable for the long and much acuminated capsules,
and the strong margin and costa to the narrow leaves.

Fig. 1. Portion of a leaf :—*magnified.*

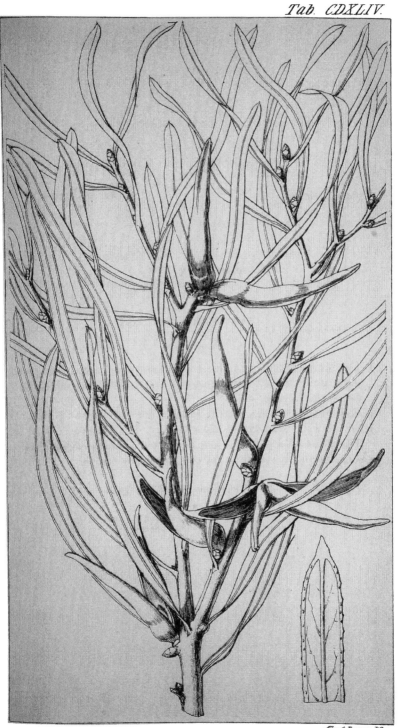

TAB. CDXLV.

HAKEA INTERMEDIA. *n. sp.*

Foliis circumscriptione ovali-oblongis basi cuneatis marginibus grosse spinoso-dentatis nitidiusculis obscure penninerviis, ramis ferrugineo-tomentosis, capsulis ovatis acuminatis gibbosis apice compressis bicalcaratis intus lævibus.

HAB. King George's Sound. *Mr. Baxter.*

A copiously branched plant with crowded foliage. It appears to be intermediate between *H. ilicifolia* and *H. nitida*, Br., having the downy branches of the former, and the fruit, internally smooth, like the latter.

TAB. CDXLVI.

XYLOMELON OCCIDENTALE. *Br.*

Foliis subellipticis, inferioribus rami floriferi passim dentatis, superioribus integerrimis, paginis omnium subsimilibus opacis utriusque epidermide glandulifera, perianthiis extus rachique tomento appresso incanis, stylo floris hermaphroditi longitudinaliter lanato. *Br.*

Xylomelon occidentale. *Br. Prodr. Suppl. p.* 31.

HAB. Baie de Géographe, South-western shores of New Holland. *Mr. Fraser.* Swan River Colony. *Mr. Jas. Drummond.*

This is a second species of *Xylomelon*, described by Mr. Brown; the original *X. pyriforme* seems to be confined to the Eastern Coast.

g

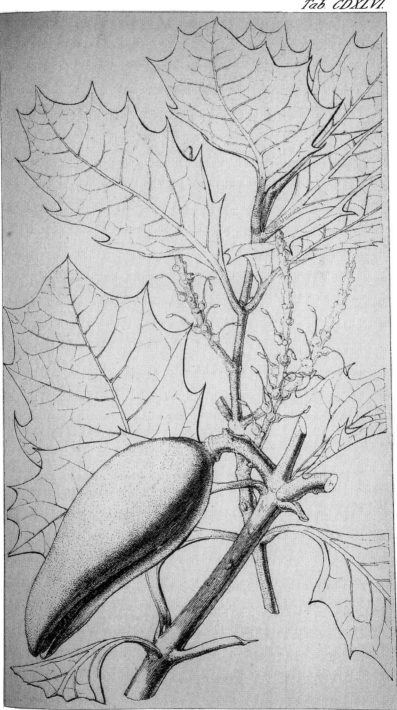

TAB. CDXLVII.

HAKEA UNDULATA. *Br.*

Foliis obovatis tri-(v. septem-) nervibus reticulato-venosis undulatis spinoso-dentatis, capsulis ecalcaratis ventricosis, (floribus minutis glaberrimis). *Br.*

Hakea undulata. *Br. Prodr. p.* 384.

HAB. New Holland, South coast. *Brown.* King George's Sound. *Mr. Baxter.*

Whole plant glabrous. Besides the three principal nerves, there are 2 or more frequently 4 others, which are parallel with them, not indeed equally originating at the base, but giving the foliage the appearance of being, at first sight, rather 7-than 3-nerved. The flowers are very small, and when dry become black.

Fig. 1. Small portion of a flowering branch; *nat. size. f.* 2. Flower scarcely expanded. *f.* 3. Flower fully expanded:— *magnified.*

TAB. CDXLVIII.

CAREX FILIFOLIA. *Nutt.*

Dioica, spica solitaria simplici superne attenuata; masc. squamis late ovatis obtusissimis lateribus involutis; fæm. squamis latissimis scariosis truncatis involutis fructum subæquantibus, fructibus ovatis obtusissime triangulatis apiculatis, seta hypogyna stricta fructu breviore, stigmatibus 3.

Carex filifolia. *Nutt. Gen. Am.* 2. *p.* 204. *Dewey Caricogr. in Sill. Journ. v.* 11. *p.* 150, *and v.* 12. *p.* 296. *tab. P. f.* 50. *Gray, N. Am. Cyp. p.* 405. *Schwein. et Torr. Car. in Ann. Lyc. N. York, v.* 1. *p.* 298. *Br. in Rich. App. Frankl. Journ. ed.* 2. *p.* 35. *Boott, in Hook. Fl. Ber. Am. v.* 2. *p.* 208.

Kobresia globularis. *Dewey Caricogr. l. c. v.* 29. *p.* 253.

Uncinia breviseta. *Gray, N. Am. Cyp. p.* 428.

HAB. Dry plains and gravelly hills of the Missouri. *Nuttall. Bradbury (in Herb. Hook.)* Woody country of Arctic America. *Dr. Richardson.* Rocky mountains. *Drummond.*

In habit allied to our well-known *Carex dioica,* but extremely different in the scales and fruit. In this species, too, there is an hypogynous seta, (though short and not uncinate), as in the genus *Uncinia,* so that it has perhaps as strong a claim to be placed as by Dr. Asa Gray in that genus, as in *Carex.* That able Botanist had not the opportunity of seeing authentic specimens of Nuttall's *C. filifolia,* and he considered Dr. Richardson's specimens distinct. But there can be no question of their identity.

Fig. 1. Male flower, with the scale, inner view. *f.* 2. Female flower, with the scale, outer view. *f.* 3. Inner view of ditto. *f.* 4. Fruit. *f.* 5. Achenium, with the hypogynous scale :— all *magnified.*

TAB. CDXLIX.

PHYSURUS VAGINATUS. *n. sp.*

Caule elongato folioso, foliis remotis oblongo-ovatis petiolatis,
petiolo basi membranaceo inflato vaginato, spica terminali
oblonga densiflora glabra, bracteis ovatis acuminatis, sepalis
petalisque oblongis labello trilobo lobo medio ovato-acuminato
apice reflexo, cornu libero ventricoso sepalis breviore.

HAB. Guatemala. *G. U. Skinner, Esq.*

Radix fibrosa, fibris villosis crassiusculis.   Caulis erectus,
spithamæus, fere ad pedalem, foliosus.   Folia remota,
oblongo-ovata, acuminata, tenui-membranacea, petiolata, 5-
9-nervia, nervis venulis connexis, petiolis brevibus basi
insigniter dilatata, vaginata, inflata, tenuissime membranacea,
hyalina, striata.   Spica terminalis, oblonga, multiflora.
Bracteæ, inferiores sæpe vacuæ, late ovatæ, acuminatæ,
hyalino-membranaceæ, longitudine ovarii.   Flores glaberrimi;
sepala oblonga, dorsale cum petalis oblongis agglutinatum.
Labellum perianthio brevius, basi calcaratum, trilobum, lobis
lateralibus rotundatis, intermedio majore, ovato, acuminato,
acumine recurvo.   Calcar labello brevius, liberum, apice
incrassatum.   Columna brevis, anthera rostelloque ovatis
acutis.

Fig. 1. Side view of a flower and bractea. *f.* 2. Front view
of ditto. *f.* 3. Upper, and *f.* 4, under side of the labellum,
(the spur being removed). *f.* 5. Column. *f.* 6. Rostellum and
anther. *f.* 7. Pollen-masses:—all *magnified.*

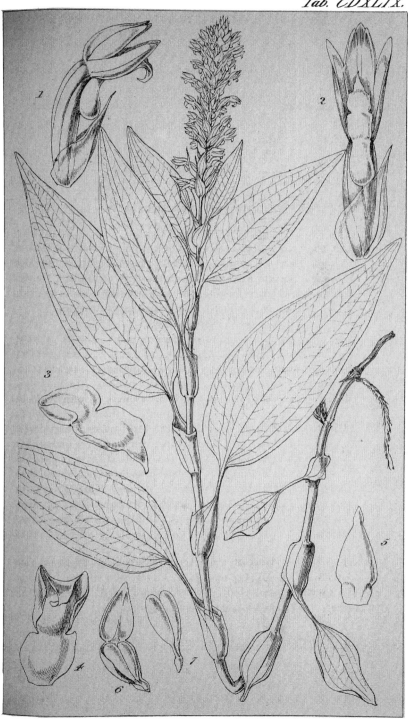

TAB. CDL.

FUCHSIA CORDIFOLIA. (*Benth.*) β!

Caule glabro, foliis oppositis v. ternatim verticillatis longe petiolatis late cordatis (ovatisve) denticulatis minute puberulis subtus subglabris, pedicellis axillaribus unifloris folio brevioribus, calycis pubescentis longe tubulosi laciniis petala ovata brevissime acuminata subduplo superantibus. *Benth.*

Fuchsia cordifolia. *Benth. Pl. Hartweg. p.* 74. *n.* 528. *Lindl. Bot. Reg.* 1841. *t.* 70.

β. foliis ovatis. (TAB. NOSTR. CDL.)

HAB. Guatemala. *G. U. Skinner, Esq.* On Zetuch, a volcano in the same country, at an elevation of 10,000 feet above the level of the sea. *Hartweg.*

It is so long since I had the impressions printed of the plate of this fine species of *Fuchsia* from Mr. Skinner's specimen, that it has now been introduced to our gardens, and has recently been published, both from Hartweg's dried specimens and from those that have flowered in our green houses. Our plant indeed does not deserve the name of *cordifolia*, the leaves being decidedly ovate, not heart-shaped, whence I have thought it better to consider this a variety.

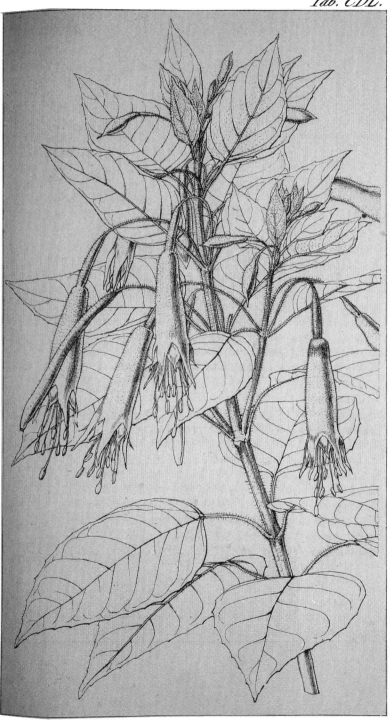

TABS. CDLI. CDLII.

Sinclairia discolor, *Hook. et Arn.*

Sinclairia, *Hook. et Arn.*—Gen. Char. *Capitulum* multiflorum radiatum : *fl.* radii ligulatis, fœmineis ; *disci* hermaphr. 5-fidis, lobis linearibus, æqualibus, demum revolutis, apicibus hirsutulis. *Receptaculum* nudum. *Involucrum* campanulatum, squamis imbricatis appressis, interioribus brevibus ovatis. *Antheræ* disci ecaudatæ, filamentis levibus. *Styli rami* elongati, fere subulati ; *disci* breviores, lobos corollæ vix superantes, subhispiduli, obtusiusculi. *Achenium* breve, glabrum, angulatum. *Pappus* fulvus, biserialis ; *serie externâ* paleaceâ, brevi ; *internâ* elongatâ, setiformi, scabrâ, rigidâ, fragili.—Frutex *glaber (vel* arbor ?) *Mexicanus.* Rami *fere ad apices lignosi.* Folia *opposita, longe petiolata, integerrima, rhomboidea, brevi-acuminata, trinervia, supra viridia, subtus albissima, nervis atro-fuscis.* Petioli *graciles, basi dilatati, amplexantes.* Panicula *terminalis, thyrsoidea, speciosa.* Flores *lutei.*

Sinclairia *discolor, Hook. et Arn. in Bot. of Beech. Voy. p.* 433. Hab. Realejo, Guatemala, on the shores of the Pacific. *Dr. Sinclair.*

In the Botany of the voyage of Capt. Beechey, Mr. Arnott and myself dedicated this plant, which we consider an entirely new genus, to our excellent friend Dr. Sinclair, who, in the surveying voyage of H. M. S. Sulphur, on the Pacific side of S. America, employed his leisure in collecting the vegetable productions of the countries he visited. We place *Sinclairia* among the *Vernoniaceæ*, near the genera *Hectoria* and *Andromachia.* The flowers are nearly an inch in diameter ; leaves 4-5 inches long, and almost as much broad, beneath quite white (but neither tomentose nor farinose,) beautifully marked with the dark brown nerves.

Fig. 1. Capitulum, *f.* 2. Floret from the disk. *f.* 3. Portion of the external series of the pappus. *f.* 4. Floret of the ray. *f.* 5. Hair from the inner series of the pappus. *f.* 6. Upper part of a corolla of the disk laid open to show the stamens :— *all magnified.*

h

Tab. CCLI. CCLII.

TABS. CDLIII. CDLIV.

Etaballia Guianensis.

Gen. Char. *Calyx* tubulosus, apice breviter 5-dentatus, subbilabiatus. *Petala* 5, ad basin calycis inserta, longissime linearia, æstivatione inflexa, imbricata. *Stamina* 10, monadelpha, alterna breviora. *Antheræ* ovatæ. *Ovarium* sessile, villosum, 2-3-ovulatum. *Stylus* brevis. *Stigma* oblique capitatum. *Legumen* ?—Arbor *ramis ramosissimis glabris.* Folia *simplicia (unifoliolata) brevissime petiolata, ovata v. ovato-oblonga breviter et acute acuminata, penninervia, coriacea, glabra v. subtus ad venas sparse pubescentia.* Spicæ *florum axillares et terminales densæ.* Bracteæ *ovato-orbiculatæ, concavæ, ante anthesin imbricatæ.* Bracteolæ *minimæ, lanceolatæ.* Flores *sessiles.* Calyx *ferrugineus.* Petala *lutea, omnia inter se subsimilia.* Stamina *calycem æquantia, ultra medium symmetrice monadelpha, tubo integro.*

Etaballia *Guianensis. Benth. in Hook. Journ. Bot. 2. p.* 99.

Hab. Abundant at the cataracts of Etabally on the Essequibo river, where it forms a strikingly beautiful tree, almost covered with bright yellow flowers, and is called by the natives *Etabally,* after the name of the cataract. *Schomburgk.*

This is a highly singular plant; being one of the very few *Leguminosæ* which cannot be recognised as belonging to that Order at first sight. It has indeed very much the aspect of an *Inocarpus*; although, on examining the structure of the flowers, it is found to be closely allied to *Schnella* (a genus including most of the small-flowered American *Bauhinieæ.*) The simple foliage without any tendency to bifurcation of the midrib is rare; but is met with in a few other species of the *Bauhinieæ.*

The supposed second species, mentioned in the work above quoted, under the name of *E. macrophylla,* must be suppressed, having originated in a mistake.

The drawing was made by Dr. Joseph Hooker, of H. M. surveying ship *Erebus. Bentham.*

Fig. 1. Flower. *f.* 2. Stamens. *f.* 3. Staminal tube cut open, showing the ovary. *f.* 4. Section of the ovary :—*all magnified.*

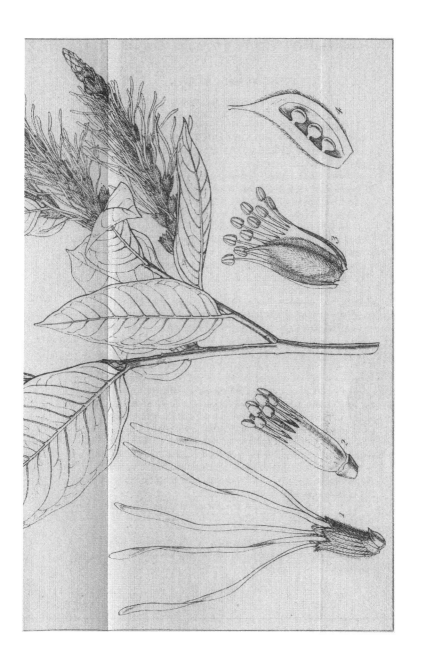

TAB. CDLV.

OCIMUM BRACTEOSUM.

Caule herbaceo erecto piloso-hispido, foliis breviter petiolatis oblongo-lanceolatis acutiusculis remote serratis basi angustatis supra glabriusculis subtus hispidulis, floralibus bracteæformibus, calyce 2-3-plo longioribus coloratis, calyce fructifero reflexo ovato subinflato dente supremo ovato breviter decurrente, lateralibus ovatis breviter mucronatis, infimis longe subulatis, filamentis edentulis.

Ocimum bracteosum. *Benth. Lab. Gen. et Sp. p.* 14.

HAB. In the fields of Lambsar in Senegambia. *Leprieur and Perottet.*

This and the nine following plates illustrate some of the genera of *Ocimoideæ*, a tribe of *Labiatæ* consisting chiefly of tropical species, and readily distinguished by their stamina, which instead of ascending under the upper lip of the corolla in pairs as in most *Labiatæ*, or spreading in all directions as in *Menthoideæ*, are turned downwards, and lie on the lower lip; a circumstance which induced the older authors to consider the flowers as resupinate. The anthers, moreover, sooner or later after they have shed their pollen, open out into an orbicular or reniform apparently unilocular disk, the two cells being always confluent. The genus *Ocimum*, as now limited, is distinguished from others of the tribe by the decurrent margins of the upper tooth of the calyx, the flat lower lip of the corolla, and from *Orthosiphon* by the style bifid at the apex with pointed lobes and minute or marginal stigmatic surfaces. *O. bracteosum* belongs to the section *Gymnocimum*, in which the filaments are entirely without appendages at the base. *Bentham.*

Fig. 1. Flower. *f.* 2. The same cut open. *f.* 3. Calyx at the maturity of the fruit. *f.* 4. Upper portion of the style. *f.* 5. Anthers. *f.* 6. Carpel. *f.* 7. Seed :—*all magnified.*

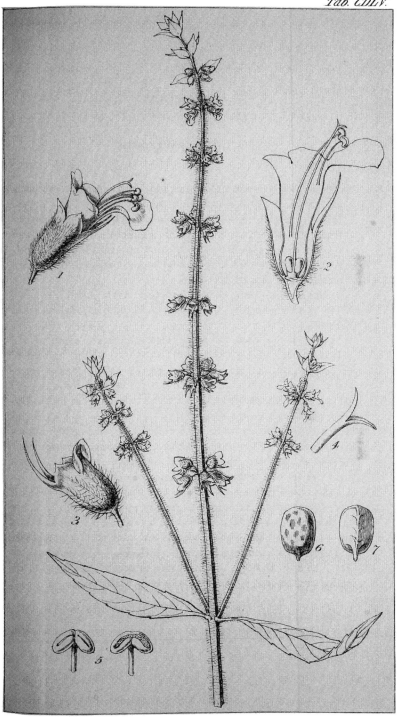

TAB. CDLVI.

ACROCEPHALUS CAPITATUS.

Caule procumbente foliisque ovatis subglabris, calycis labio inferiore 4-dentato.

Prunella indica. *Burm. Fl. Ind. p.* 130.
Ocimum capitellatum. *Linn. Mant. p.* 276.
Ocimum capitatum. *Roth, Nov. Pl. Sp.* 276.
Acrocephalus capitatus. *Benth. Lab. Gen. et Sp. p.* 23.

HAB. Common in moist situations over the greater portion of East India, in the Burman Empire, in Java, and, according to *Willdenow,* in China.

This little plant has much the appearance of an *Escholtzia* and in some respects approaches that genus in character. The decidedly declinate stamens have, however, placed it amongst *Ocimoideæ,* where, with a Javanese plant (probably a mere variety) and a Madagascar species, distinguished by the entire lower lip of the calyx, it forms a genus differing from *Ocimum,* *Geniosporum* and *Moschosma* in the form of the calyx, and more especially in inflorescence, and from all other *Ocimoideæ* by the all but regular corolla. The calyx is tubular, as in several *Geniospora;* but in the latter genus the lateral teeth are more or less connected with the upper one into an upper lip, whilst in *Acrocephalus* the four lower teeth form the lower lip, leaving the upper tooth solitary. *Bentham.*

Fig. 1. Flower. *f.* 2. Ripe calyx. *f.* 3. Corolla cut open. *f.* 4. Upper portion of the style. *f.* 5. Bract. *f.* 6. Carpel :— *all magnified.*

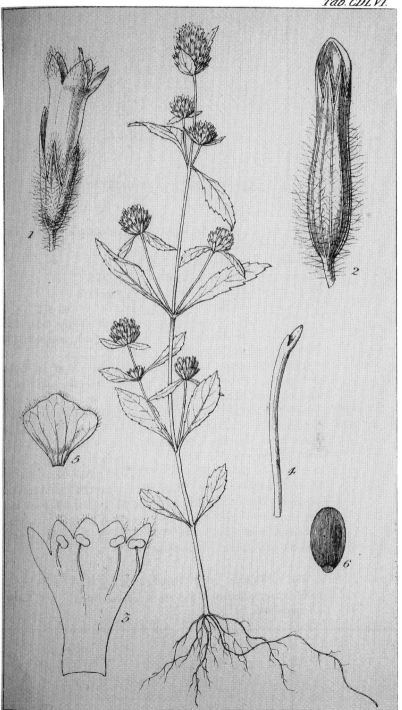

TAB. CDLVII.

MARSYPIANTHES HYPTOIDES.

Hyptis Chamædrys. *Willd. Sp. Pl.* 3. *p.* 85. *Poit. Ann. Mus. Par.* 7. *p.* 468.

H. pseudochamædrys. *Poit. Ann. Mus. Par.* 7. 469.

H. inflata. *Spreng. Syst.* 2. *p.* 731.

H. lurida. *Spreng. l. c.*

Marsypianthes hyptoides. *Mart. in Benth. Lab. Gen. et Sp. p.* 64.

HAB. A very common weed, especially near the sea, in the greater part of tropical America, from Mexico to Guayaquil on one coast, and to South Brazil on the other.

This species varies much in aspect, but the different forms can hardly be considered as distinct species. It constitutes alone a genus, with the habit and general character of the capitate *Hyptides*, but differing from them in the broadly campanulate calyx, and especially in the very singular form of the carpels, the margins of which are expanded into a membranous wing, with the edges toothed and bent inwards, so as to give to the whole carpel a kind of boat shape. The flower is precisely that of a *Hyptis. Bentham.*

Fig. 1. Flower. *f.* 2. Corolla cut open. *f.* 3, 4. Anthers. *f.* 5. Upper portion of the style. *f.* 6. Mature calyx. *f.* 7. Fruit, as enclosed in the calyx. *f.* 8. Single carpel viewed from behind. *f.* 9. The same seen in front. *f.* 10. Section of the same:—*all magnified.*

TAB. CDLVIII.

HYPTIS VERTICILLATA.

Suffruticosa, ramis erectis pubescentibus v. pilosis, foliis breviter petiolatis lanceolatis acutis serratis basi rotundatoangustatis tenuissime pubescentibus, verticillastris laxiusculis plurifloris distinctis racemosis, calycis ovati glabri dentibus erectis ovato-lanceolatis.

Hyptis verticillata. *Jacq. Ic. Rar.* 1. *t.* 113. *Benth. Lab. p.* 130.

Mentha hyptiformis. *Lam. Dict.* 4. *p.* 110.

Stachys patens. *Swartz.*

HAB. Common on the roadsides, in various parts of the warmer regions of Mexico, in St. Domingo, and perhaps some other of the West Indian Islands.

The genus *Hyptis*, together with the small allied genera, *Peltodon, Marsypianthes,* and *Eriope,* consists entirely of American species, and is readily known among *Ocimoideæ* by the pouch-shaped hanging lower division of the corolla, attached by so narrow a base that it appears often almost articulate. It is one of the most extensive in the Order, as there are above 220 species known; most of them natives of the lower mountainous regions of South America, and a few of them exceedingly common wherever cultivation has commenced under the tropics in the new world and even in the old world, where they have probably been introduced by man. There is a very great diversity in habit, but little in structure of the flower, in the different species which have been distributed into nineteen sections founded chiefly on inflorescence. The *H. verticillata* belongs to the fifteenth section *Minthidium,* consisting of herbs or undershrubs, with the flower-cymes sessile or nearly so, many-flowered, and condensed into verticillasters as in the majority of *Labiatæ,* the calyx regular, the corolla scarcely protruding from it, and the bracts inconspicuous. The species have thus very much the appearance of *Menthæ,* in everything but the corolla and stamens. *Bentham.*

Fig. 1. Flower. *f.* 2. The same cut open. *f.* 3, 4. Anthers. *f.* 5. Upper portion of the style. *f.* 6, 7. Carpels :—*all magnified.*

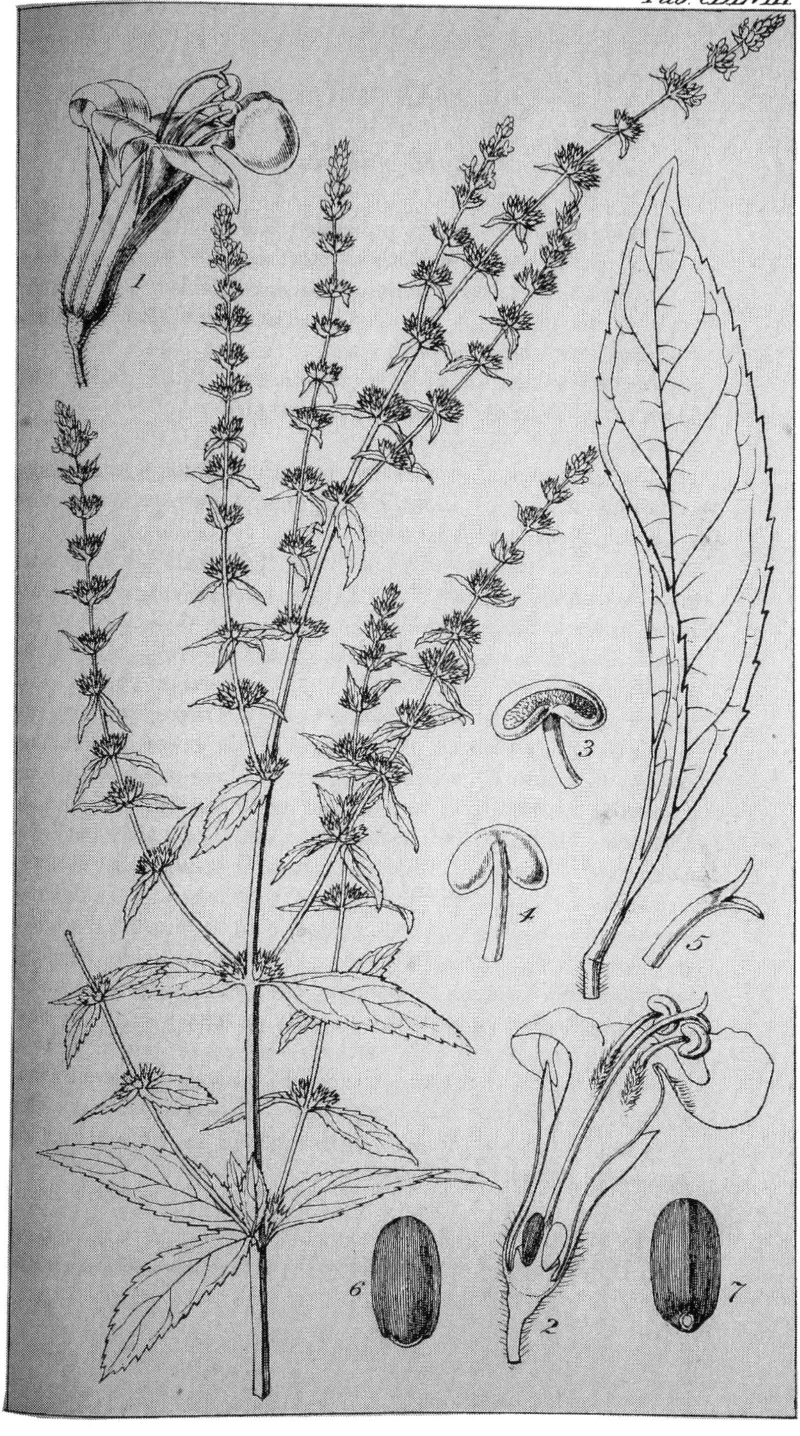
1
2
3
4
5
6
7

TAB. CDLIX.

ORTHOSIPHON RUBICUNDUS.

Caulibus cæspitosis basi foliosis ramosis, foliis oblongo-ovatis grosse dentatis basi angustatis infimis petiolatis, superioribus sessilibus, corollæ tubo rectiusculo, calyce duplo longiore, fauce subæquali, staminibus corollâ parum brevioribus.

Orthosiphon rubicundus. *Benth. Lab. Gen. et Sp. p.* 26.

Plectranthes rubicunda. *Don, Prod. Fl. Nep. p.* 116.

Lumnitzera rubicunda. *Spreng. Syst. Cur. Post. p. 223.*

HAB. Along the mountainous regions of North India, from the Kheesee Pass? at the entrance of Deyra Dhoun (*Royle*), to the Burmese territory. *Wallich.*

The genus *Orthosiphon* has much of the habit, the calyx, and most of the characters of *Ocimum*; but the tube of the corolla is usually longer, and the apex of the style, instead of being divided into two linear pointed lobes, is almost entire and capitate, with a terminal stigmatic surface. The species are all Asiatic or African, excepting a remarkable one, contained in the South American herbarium transmitted by Pavon to the late Mr. Lambert, but of which the precise station is as yet unknown. *Bentham.*

Fig. 1. Flower. *f.* 2. Corolla cut open. *f.* 3, 4. Anthers. *f.* 5. Upper portion of the style (represented too much flattened.) *f.* 6. Mature calyx. *f.* 7. Fruit. *f.* 8. Single carpel. *f.* 9. Floral leaf:—*all magnified.*

TAB. CDLX.

PLECTRANTHUS TERNIFOLIUS.

Tomentoso-villosus, caule erecto subramoso, foliis ternatim verticillatis subsessilibus lanceolato-oblongis acuminatis serratis basi cuneatis rugosis, paniculis ramosis densis pyramidatis multifloris, calycibus fructiferis cylindricis erectis striatis æqualiter 5-dentatis.

Plectranthus ternifolius. *Don, Prod. Fl. Nep.* 117. *Benth. Lab.* 44.

Ocimum ternifolium. *Spreng. Syst. Cur. Post. p.* 224.

HAB. On the roadsides, in the damp wooded regions along the Himalaya, from Kamaon to the Burmese territory.

Plectranthus, one of the largest genera of Asiatic *Ocimoideœ,* is distinguished from *Ocimum* by the concave lower division of its corolla, and from *Coleus* by the stamens not connected into a tube. It varies in habit and calyx, as well as in the form of the tube of the corolla, from which characters it has been divided into seven sections. To these ought perhaps to be added three more, *Anisochilus, Æollanthus* and *Pycnostachys,* genera which have been founded merely upon peculiarities in the form of the calyx. The *P. ternifolius,* along with a closely allied South African species, forms the section *Pyramidium;* characterised by an erect, tubular or ovate, equally 5-toothed calyx (in the fruit-bearing state), a straight corolline tube, and a dense pyramidically paniculate inflorescence. *Bentham.*

Fig. 1. Flower. *f.* 2. Corolla cut open. *f.* 3. Stamen. *f.* 4. Anther seen from the back. *f.* 5. Upper portion of the style. *f.* 6. Ovary. *f.* 7. Single carpel seen from the side:— *all magnified.*

TAB. CDLXI.

ERIOPE MACROSTACHYA.

Fruticosa, ramis pubescentibus villosisve, foliis petiolatis ovato-
lanceolatis acutis denticulatis basi rotundatis subcordatisve
rarius cuneatis rugosis utrinque villosis, panicula ampla ramosa.

Eriope macrostachya. *Mart. in Benth. Lab. Gen. et Sp. p.* 145.
HAB. Elevated *Campos,* and woods of the mining districts in
Brazil. *Martius and others.*

The essential character, derived from the corolla, is very
nearly the same in *Eriope* as in *Hyptis,* and the affinity with the
section *Hypenia* of that genus is certainly very close. Yet the
peculiar form of the mature calyx, bilabiate and closed at the
mouth with hairs, appears constant; as is also the inflorescence,
the flowers being solitary and opposite as in *Scutellaria,* forming
leafless simple or paniculately branched racemes. There are
about fifteen species known, all Brazilian. *Bentham.*

Fig. 1. Flower. *f.* 2. Calyx cut open. *f.* 3. Corolla cut open.
f. 4, 5. Stamens. *f.* 6. Ovary and style :—*all magnified.*

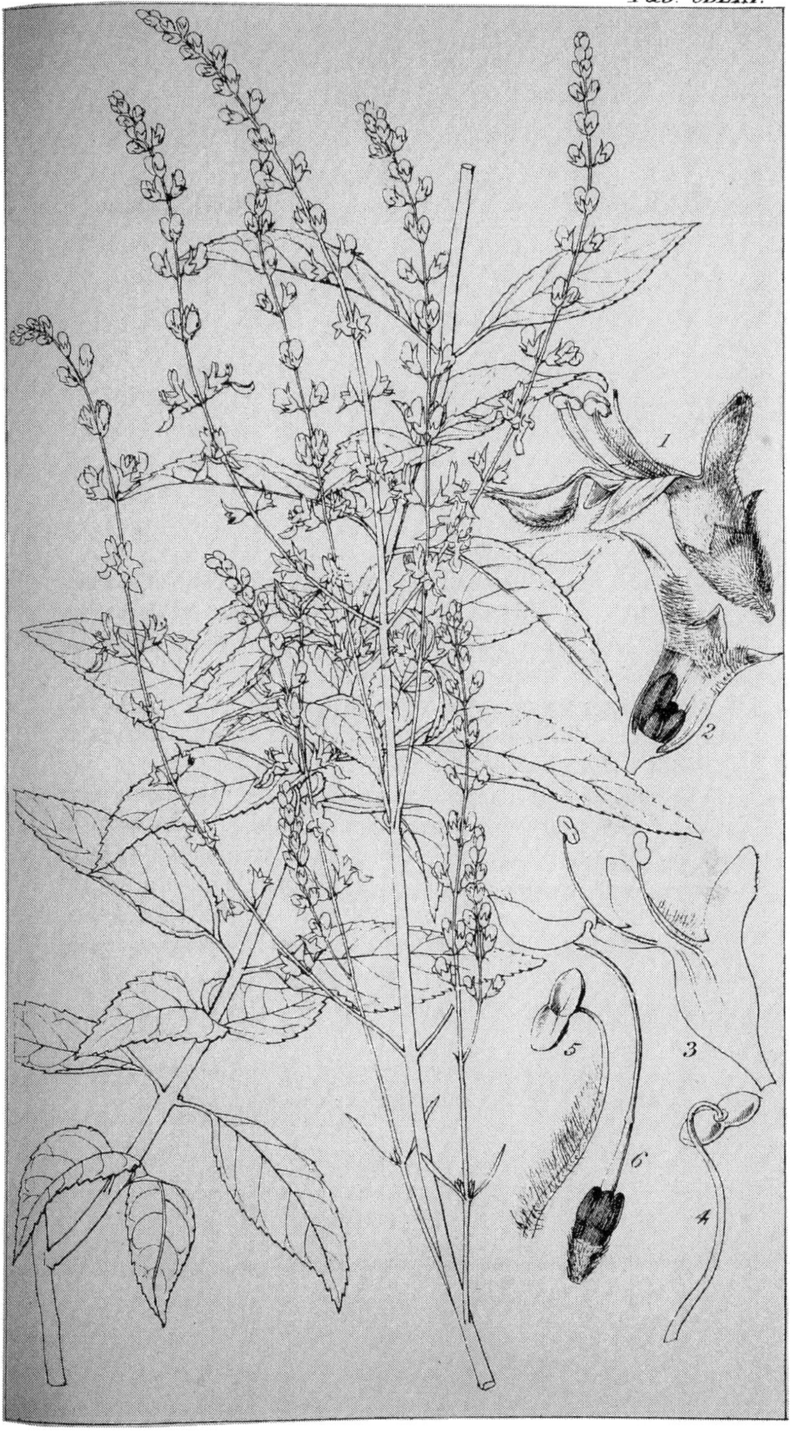
1
2
3
4
5
6

TAB. CDLXII.

GENIOSPORUM STROBILIFERUM.

Caule erecto ramoso, foliis subsessilibus ovato-oblongis v. ovato-lanceolatis utrinque angustatis supra hispidulis subtus glabriusculis, verticillastris multifloris in apice ramorum spicatis infimis subremotis, foliis floralibus ovatis acuminatis flores superantibus, calycibus subsessilibus, fructiferis erectis striatis basi transverse rugosis, ore membranaceo irregulariter 5-dentato.

Geniosporum strobiliferum. *Wall. Pl. As. Rar.* 2. *p.* 18. *Benth. Lab. p.* 20.

HAB. In North India, along the whole range of the Himalaya.

The corolla of *Geniosporum* is the same as that of *Ocimum* and *Moschosma*, but the upper lobe of the calyx is not large and decurrent as in *Ocimum*, and *Moschosma* has a clavate style. The habit of *Geniosporum* is different from that of any of the allied genera. The verticillasters are dense and many-flowered, the upper floral leaves and summits of the calyces are frequently white or coloured, and the ripe calyx is usually marked with transverse reticulations at its base. *Bentham.*

Fig. 1. Flower. *f.* 2. Corolla cut open. *f.* 3. Anthers. *f.* 4. Upper portion of the style. *f.* 5. Mature calyx. *f.* 6. Fruit. *f.* 7. Single carpel:—*all magnified.*

TAB. CDLXIII.

HYPTIS SALZMANNI.

Fruticosa, ramis foliatis patentim pilosis, foliis petiolatis ovatis obtusis eroso-crenatis rugosis pubescentibus subtus pallidis, panicula laxissima subnuda glaberrima glauca, ramis elongatis, pedunculis filiformibus 1-3-floris, calycibus campanulatis venosis, dentibus æqualibus acutis, corollæ tubo calyce subduplo longiore.

Hyptis Salzmanni. *Benth. Lab. Gen. et Sp. p.* 138.

HAB. Along the Rio San Francisco, from the province of Minas Geraes to its mouth, and in various parts of the province of Bahia.

This species belongs to the section *Hypenia,* remarkable for its peculiar habit, the lower portion of the plant being invariably clothed with long spreading hairs, whilst the panicle is always perfectly smooth, and more or less glaucous. The inflorescence approaches that of *Eriope,* and in some species the great length of the tube of the corolla alters much the appearance of the flower; yet these characters are so ill-defined and connected by so many intermediate states with more ordinary forms of *Hyptis,* that it would be highly inconvenient to adopt them as generic distinctions. Many of the species are very handsome, with scarlet flowers above an inch in length. *Bentham.*

Fig. 1. Flower. *f.* 2. Mature calyx. *f.* 3. Corolla cut open. *f.* 4. Anther. *f.* 5. Upper portion of the style. *f.* 6. Fruit. *f.* 7. Single carpel :—*all magnified.*

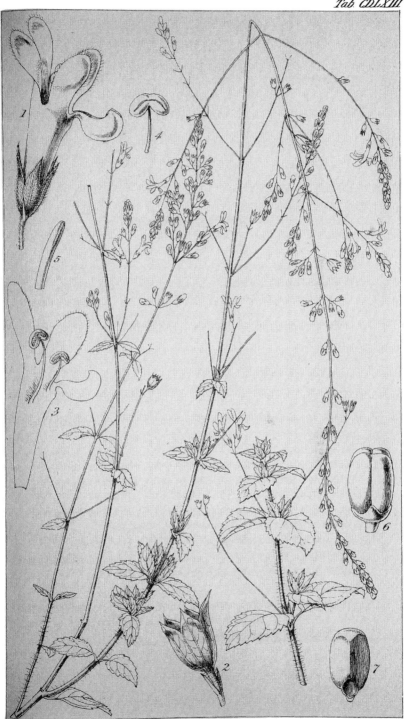

TAB. CDLXIV.

Plectranthus scrophularioides.

Caule herbaceo erecto ramoso subglabro, foliis longe petiolatis lato-ovatis crenatis basi rotundatis inæqualiter cordatis v. subcuneatis, floralibus bracteisque minutis, paniculis laxis, calycibus fructiferis declinatis profunde bilabiatis inflatis, labio superiore adscendente tridentato inferiore concavo porrecto breviter bidentato, dentibus omnibus obtusis, corollis inflatis supra gibbis calyce subtriplo longioribus, staminibus exsertis. Plectranthus scrophularioides. *Wall. Pl. As. Rar.* 2. *p.* 16. *Benth. Lab. p.* 40.

Hab. North India, along torrents in Nepal and Kamaon. *Wallich.*

The section of *Plectranthus*, to which this plant belongs, was established by Schrader as a genus, under the name of *Isodon*; the teeth of the calyx, in the species which he described, being nearly equal and scarcely bilabiate, even at maturity. The name, having been thus applied, was adopted for the section, although not so suitable to the majority of its species, in which the calyx is more or less decidedly bilabiate. In the *P. scrophularioides* it is deeply so. The true character of the section consists in the lateral teeth of the calyx being more or less connected with the upper one, not with the lower ones as in *Coleoides*, in the want of that spur to the corolla which distinguishes *Germanea* and *Melissoides*, and the ripe calyx being declinate, not erect as in *Pyramidium* and *Amethystoides. Bentham.*

Fig. 1. Flower. *f.* 2. Corolla cut open. *f.* 3. Mature calyx. *f.* 4. Anther. *f.* 5. Upper portion of the style. *f.* 6. Fruit. *f.* 7. Single carpel :—*all magnified.*

TAB. CDLXV.

ILEX AFFINIS. *Gardn.*

Glaberrima, foliis oblongo-lanceolatis utrinque attenuatis supra
    medium obtuse et distanter serratis inferne integerrimis,
    racemis 2-3 axillaribus paniculatis densifloris, calyce glabro.
Ilex affinis. *Gardn. Herb. Bras. n.* 3086.
HAB. In wooded ravines in the Serra de Natividade, province
of Goyaz, Brazil. January, 1840.
This species is nearly related to the *Ilex Paraguayensis*, (see
Journ. of Bot. Tab. I. and II.), but is readily distinguished,
both in the living and dried state, by its very thick coriaceous
leaves, which are also more obtusely and distantly serrated, and
less cuneated; and by its more numerous and more densely
flowered racemes. This is the most northern species I have
met with in Brazil, and although not uncommon about the Villa
de Natividade, I have never seen its leaves collected to be made
into tea. In my Goyaz collections there is another species, with
much broader, shorter and nearly entire leaves, shorter and
fewer-flowered racemes, and with flowers nearly twice as large.
It may be characterized as follows:
Ilex *rivularis;* glaberrima, foliis obovatis obtusis versus apicem
    obscure crenato-serratis basi acutis, racemis 2-4 axillaribus
    vix petiolo duplo longioribus, pedicellis unifloris, calyce
    pubescente, drupis (siccis) 4-sulcatis.
Ilex rivularis. *Gardn. Herb. Bras. n.* 3085.
HAB. In woods by the sides of streams near Villa de Natividade,
province of Goyaz, Brazil. January, 1840.
Frutex 10-15 pedalis, glaberrimus, ramulis pauce angulatis.
Folia 4-4½ poll. longa, 2 circiter lata. *G. Gardner.*

Fig. 1, 2. Flowers. *f.* 3. Pistil, and the corolla laid open :—
magnified.

TAB. CDLXVI.

TAPURA CILIATA. *Gardn.*

Foliis oblongis obtusis versus basi subcuneatis supra glaberrimis subtus villosis margine revolutis dense villoso-ciliatis, petiolis floriferis, floribus in glomerulum dense aggregatis sessilibus.

Tapura ciliata. *Gardn. Herb. Bras. n.* 3087.

HAB. Rare in dry, open woods between the Mission of Duro, and Villa de Natividade, in the province of Goyaz, Brazil. January, 1840.

Arbor 12-16 pedalis ramosissima. *Ramuli* fusco-tomentosi. *Folia* coriacea, alterna, petiolata, oblonga, obtusa, basi subcuneata, supra glaberrima, subtus villoso-tomentosa, margine revoluta, dense ciliata. *Petioli* breves, villosi, apice floriferi. *Stipulæ* parvæ, triangulares, deciduæ. *Pedicelli* cum petiolo concreti. *Flores* flavi, in apice petioli dense aggregati, sessiles. *Calyx* basi 3-bracteatus, 5-partitus, lobis inæqualibus, ovatis, obtusis, villosis. *Corolla* gamopetala imo basi calycis concreta, tubo intus villoso, limbo subbilabiato, labio superiore 2-lobo, lobis late obovatis emarginatis, inferiore 3-lobo, lobis lineari-lanceolatis. *Stamina* 5. *Filamenta* cum petalis cohærentia, iisdem alterna et æquilonga, 3 superiora antherifera, 2 inferiora sterilia. *Antheræ* introrsæ, oblongæ, biloculares, longitudinaliter dehiscentes. *Stylus* filiformis, villosus, exsertus. *Stigma* trilobum. *Ovarium* ovato-trigonum, triloculare.

This species of *Tapura* is very distinct from that figured by Aublet at Tab. 48 of his Plant Guian., which has hitherto been the only known species of the genus. The Brazilian one is readily distinguished by its densely ciliated leaves, and the greater number of its flowers. The structure of the corolla is also different from that of the plant of Aublet. The upper lip of the latter has only one lobe, and the lower two; whereas in mine the upper lip has two broad emarginate lobes, and the lower three linear-lanceolate ones, nearly equal in length to the others. In structure the present plant is truly gamopetalous, the filaments forming the bond of union, and consequently alternating with the segments. Aublet says :—" Filamenta 5, duo ad latera labii superioris, duo breviora tubo corollæ sub labio superiori, quintum longissimum ad basin labii inferioris." Judging from what is to be seen in my plant, I should imagine that Aublet has not correctly defined the position of the stamina.— *G. Gardner.*

Fig. 1. Single flower and bracteas. *f.* 2. Corolla laid open. *f.* 3. Ovary. *f.* 4. Hypogynous gland :—*magnified.*

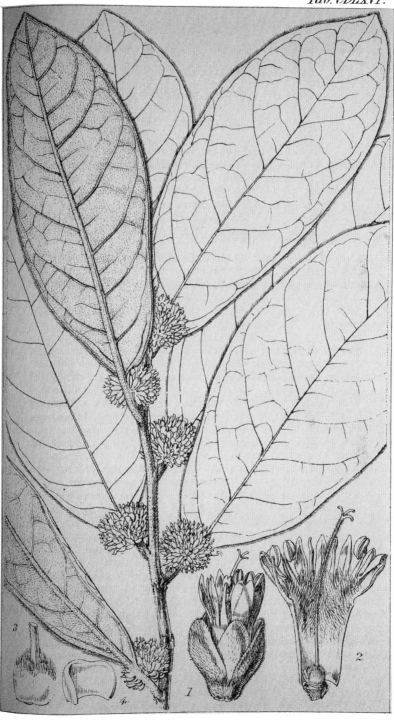

TAB. CDLXVII.

ADIANTUM CALCAREUM. *Gardn.*

Frondibus pinnatis glabris, pinnis dissimilibus, superioribus dimidiatis subtriangularibus basi truncatis margine superiore incisis, inferioribus flabellatis profunde incisis, laciniis emarginatis basi acutis vel subcordatis, indusiis lævibus, rachi glabra apice sæpe nuda elongata radicante.

Adiantum calcareum. *Gardn. Herb. Bras. n.* 3551.

HAB. In clefts of calcareous rocks near Natividade, province of Goyaz, Brazil. December, 1839.

Frondes fasciculatæ. *Stipes* subpollicaris, atropurpureus, teres, nitidus, subpaleaceus. *Rachis* teres, glabra, in apice frondis sæpe nuda, elongata, extremitate demum radicante. *Frons* 4-6 pollicaris, pinnata. *Pinnæ* fere semipollicares, alternæ, brevissime petiolatæ, superiores dimidiatæ, subtriangulares, basi truncatæ, margine superiore incisæ; inferiores flabellatæ, profunde inciso-lobatæ, laciniis emarginatis, basi acutæ vel subcordatæ. *Venæ* radiatæ, pluries furcatæ, venulis parallelis. *Sori* marginales, oblongi. *Indusia* oblonga, membranacea, glabra.

This species of *Adiantum* comes near *A. caudatum*, Linn., but differs in being a much smaller plant, thinner in texture, and smooth. The pinnæ are also shorter, broader, more deeply incised, and less recurved than they are in *A. caudatum*. The fronds of both species are occasionally radicant at their apices; and sometimes the lower pinnæ in *A. caudatum* assume the rounded flabellate form, which in the present plant proceeds half-way up the rachis. *G. Gardner.*

Fig. 1. Lower pinna. *f.* 2. Sorus; the indusium laid open:— *magnified.*

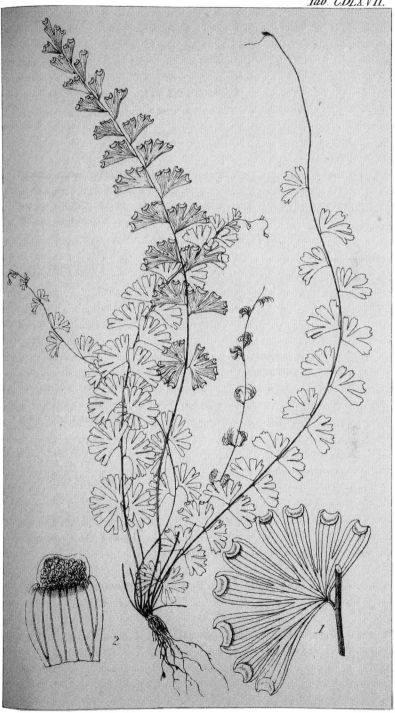

TAB. CDLXVIII.

ACHIMENES MULTIFLORA. *Gardn.*

Annua tota hirsuta erecta, foliis petiolatis oppositis ternisve ovatis acutis basi obtusis argute subduplicato-serratis, pedunculis axillaribus 3-5 floris infimis elongatis supremis subsessilibus, calycis lobis linearibus erectis dense hirsutis, corollæ tubo infundibuliformi hinc basi gibbo, lobis rotundatis.

Achimenes multiflora. *Gardn. Herb. Bras. n.* 3873.

HAB. On dry banks in woods on the Serra de Santa Brida, and near Villa de Arrayas, in the province of Goyaz, Brazil.

Herba annua, tota hirsuto-villosa, 1-1½ pedalis. *Caules* simplices. *Folia* 2½-3 poll. longa, pollicem circiter lata, opposita vel raro verticillata. *Petioli* 4-6 lineam longi. *Pedunculi* axillares, 3-5 flori. *Pedicelli* erecti, *corolla* dimidio breviores. *Calycis tubus* ovario adnatus, limbus 5-partitus, lobis linearibus obtusis. *Corolla* pallide purpurea, glabra, tubuloso-infundibuliformis, basi postice hinc gibba, limbo irregulariter bilabiato, 5-fido, lobo medio labii inferiore subdenticulato, lobis reliquis integris rotundatis. *Stamina* 4 didynama, antheris inter se cohærentibus. *Annulus* perigynus integer. *Stylus* apice bifidus, lobis latis obtusis intus stigmatiferis. *Ovarium* villosissimum.

The corolla of this pretty little plant is almost that of *Gloxinia*, but the bifid stigma and entire annulus prove it to be a species of *Achimenes*. It is probably allied to *A. hirsuta, DC.*, which is also Brazilian. *G. Gardner.*

TAB. CDLXIX.

TAPINA VILLOSA. *Gardn.*

Herbacea simplex erecta villosa, foliis ovatis vel ovato-oblongis utrinque obtusis vel acutiusculis grosse serratis supra dense pilosis subtus præcipue ad nervos villosis, pedunculis axillaribus 1-floris, calycis tubo brevi, lobis 5 lanceolatis, corollæ tubo brevi hinc basi gibbo.

Tapina villosa. *Gardn. Herb. Brasil. n.* 3875.

HAB. In dry clefts of rocks near the summit of the Serra de Natividade, in the north of the province of Goyaz, Brazil, February, 1840.

Herba pusilla, 1-5 uncialis, tota villosa, villi articulati. *Radix* carnosa, squamosa, fibrosa, fibrillis villosis, fuscis. *Caules* solitarii, simplices. *Folia* 1-1½ poll. longa, 8-9 lin. lata. *Pedunculi* axillares, solitarii, uniflori, internodo longiores. *Calyx* liber, 5-partitus, lobis subæqualibus, lanceolatis. *Corolla* hypogyna, infundibuliformis, tubo brevi purpurascente, basi postice gibbo, limbo albo, quinquefido, subæqualiter patente, lobis obtusis. *Stamina* 4, didynama, cum quinto rudimentario. *Filamenta* glabra. *Antheræ* ovatæ, basi cordatæ, cohærentes. *Annulus* hypogynus integer, postice in glandulam tumens. *Ovarium* ovatum, villosum. *Stylus* simplex, apice subincrassatus. *Stigma* capitato-bilobum. *Fructus* non vidi.

This little plant agrees with the characters of *Tapina* in every thing except the form of the corolla, which has a shorter and more regular tube than the species described and figured by Martius in the Nov. Gen. et Sp. Plantarum. In one of the flowers which I examined of this plant, I found the fifth filament bearing a perfect anther, which cohered with the other four.— *G. Gardner.*

Fig. 1, 2. Specimens of Tapina villosa :—*nat. size.*

k

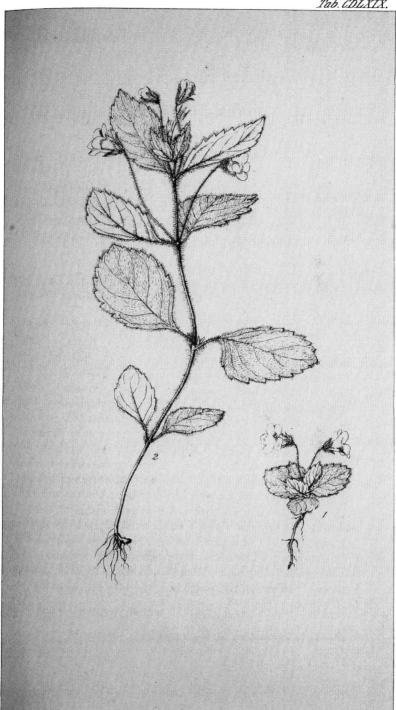

TAB. CDLXX.

ECHITES PULCHELLA. *Gardn.*

Suffruticosa, erecta, glaberrima, foliis oblongo-lanceolatis acutis
basi subcuneatis, pedunculis valde elongatis 4-6-floris, calycis
laciniis subulatis.

Echites pulchella. *Gardn. Herb. Bras. n.* 3886.

HAB. In a moist upland *campo* near Villa de Arrayas, province
of Goyaz, Brazil. March, 1840.

This very rare species of *Echites* is a suffruticose plant,
about a foot and a half high, glabrous in all its parts. *Leaves*
2-2½ inches long, and from 4 to 6 lines broad, opposite, oblong-
lanceolate, acute, narrowed towards the base, with a slightly
thickened margin. *Flowers* 4-6 on a peduncle which is more
than half the length of the whole plant. *Pedicels* 4-6 lines
long. *Calyx* small, 5-parted: segments subulate. *Corolla*
infundibuliform, scarlet; tube about three quarters of an inch
long, contracted a little at the apex; segments oblong-lanceolate,
acute, spreading. *Stamens* inserted on the tube of the corolla
near its base. *Filaments* short, villous. *Anthers* sagittate,
cohering by their middle to the *stigma. Fruit* not seen.—*G.
Gardner.*

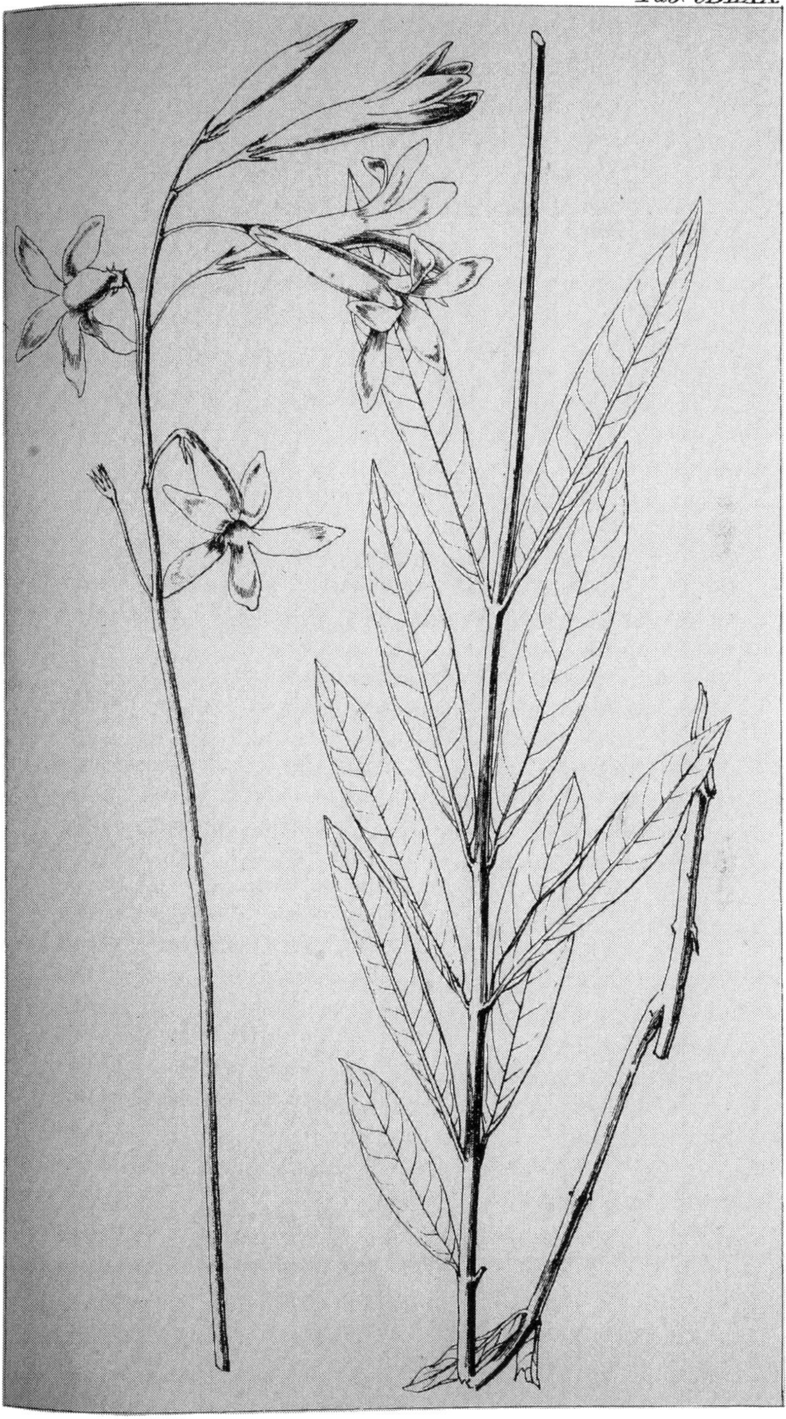

TAB. CDLXXI.

Ipomæa (Orthipomæa) neriifolia. *Gardn.*

Fruticosa, ramosa, erecta, foliis confertis vix petiolatis longe
linearibus utrinque attenuatis margine revolutis pellucido-
punctatis hirsutis, pedunculis subtrifloris, calycis piloso-
pubescentibus laciniis inæqualibus late oblongis obtusis,
corollæ tubo infundibuliformi limbo patente parum lobato.
Ipomæa neriifolia. *Gardn. Herb. Bras. n.* 3906.
Hab. Rare in dry exposed places on the Serra de Natividade,
province of Goyaz, Brazil, February 1840.
Frutex bipedalis. *Rami* teretes, striati, villoso-tomentosi. *Folia*
conferta, alterna, vix petiolata, utrinque attenuata, villosa,
pellucido-punctata, punctis rotundatis. *Pedunculi* axillares,
villosi, breves, subtriflori. *Pedicelli* pedunculo subæquales.
Calycis foliola subinæqualia, late oblonga, obtusa, concava,
piloso-pubescentia. *Corolla* pallide violacea, tubo infundibuliformi, limbo repando, patente. *Stamina* erecta.
Another species of *Ipomæa*, belonging to the same section,
which I possess from a more northern part of the province of
Goyaz, may be characterized as follows:
Ipomæa (Orthipomæa) *hirsutissima;* fruticosa, erecta, tota hirsutissima, caule simplici, foliis brevi-petiolatis oblongo-lanceolatis apice acutis cuspidatis basi rotundatis cordatisve,
pedunculis axillaribus 1-floris folio triplo brevioribus, pedicellis basi bibracteatis, bracteis magnis foliaceis lanceolatis
longe petiolatis, calycis dense hirsuti laciniis lanceolatis
acuminatis, corollæ tubo infundibuliformi extus hirsuto limbo
patente repando.
Ipomæa hirsutissima. *Gardn. Herb. Bras. n.* 3355.
Hab. In dry upland *campos,* near the Mission of Duro, province of Goyaz, Brazil. Oct. 1839.
Caules plures, vix pedales. *Corolla* roseo-violacea. *Antheræ*
erectæ, tubo inclusæ.—*G. Gardner.*

TAB. CDLXXII.

GLOXINIA ICHTHYOSTOMA. *Gardn.*

Annua, caule elongato erecto hirsuto-villoso, foliis subinæquilateris ovatis acutis basi rotundatis vel subcordatis grossè crenato-serratis utrinque hirsutiusculis, pedicellis axillaribus solitariis 1-floris, calycis 5-partiti lobis lineari-lanceolatis patentibus, corollæ tubo infundibuliformi campanulato, limbo subbilabiato, lobo intermedio labii inferioris margine incurvato longeque denticulato-ciliato.

Gloxinia ichthyostoma. *Gardn. Herb. Bras. n.* 3304.

HAB. In shady rocky places on dry calcareous hills near Arrial da Chapada, province of Goyaz, Brazil. January, 1840.

Herbacea, annua, erecta, hirsuta, 1-1½ pedalis. *Caules* simplices. *Folia* opposita, petiolata, 2½-3 poll. longa, 1½-2 pollices lata, subobliqua, grossè crenato-serrata, acuta, basi subcordata. *Petioli* 3 lineas circiter longi, dense hirsuti. *Pedicelli* solitarii, axillares, erecti, internodo triplo breviores. *Calycis tubus* ovario adnatus. *Limbus* 5-partitus, lobis lineari-lanceolatis patentibus. *Corolla* purpureo-violacea, extus pubescens, tubo infundibuliformi-campanulato basi ecalcarato, limbo subbilabiato 5-lobo, lobo intermedio labii inferioris margine incurvato longeque dentato-ciliato. *Stamina* 4, didynama et quintum rudimentarium. *Antheræ* inter se cohærentes. *Glandulæ* 5, perigynæ. *Stylus* versus apicem incrassatus. *Stigma* orbiculato-concavum.

The specific name which I have given to this species of *Gloxinia* was suggested by the very marked resemblance which the middle lobe of the lower lip of its corolla bears to the jaw of a fish. The same appearance exists, but in a slighter degree, in the original *Gloxinia maculata.—G. Gardner.*

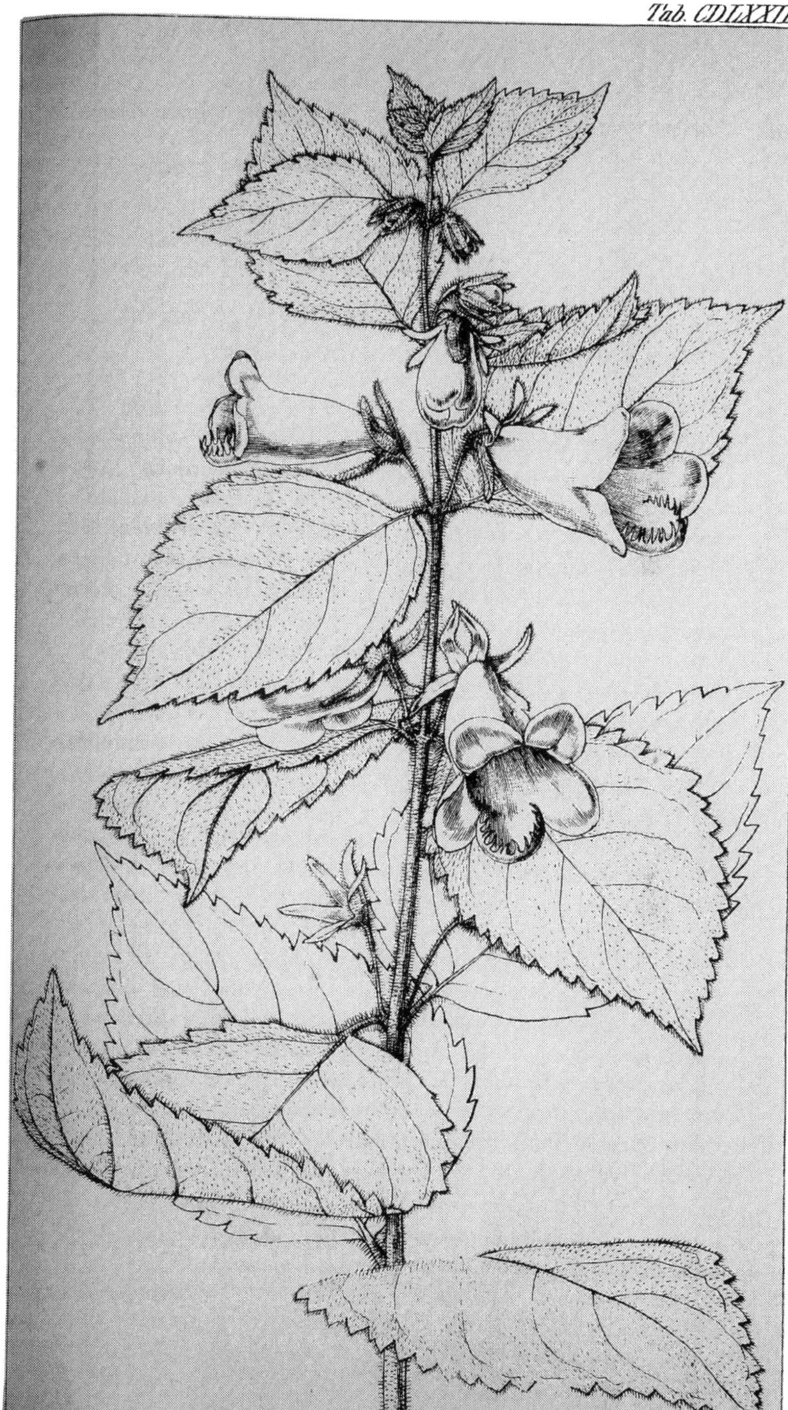

TABS. CDLXXIII. CDLXXIV.

CLEISTES SPECIOSA. *Gardn.*

Labello convoluto truncato emarginato sepalorum longitudine, lamellis infra medium integris.

Cleistes speciosa. *Gardn. Herb. Bras. n.* 4003.

HAB. Marshy places, in upland *campos* near Natividade, and between Natividade and Arrayas, province of Goyaz, Brazil, flowering from January till March.

The following description of this beautiful plant was drawn up from recent specimens.

Herbaceous, 3-4 feet high. *Root* fibrous, fibres succulent. *Stem* erect, fistular, leafy. *Leaves* between succulent and coriaceous, glaucous, finely striated with parallel veins, 5-6 inches long and about an inch and a half broad, oblong-lanceolate, their margins running down and meeting at a little more than an inch below the point where the middle part of the leaf separates from the stem. *Flowers* about 3 inches long, rose-coloured, solitary in the axils of the two or three upper leaves. *Sepals* patent, linear-lanceolate, acute. *Petals* conniving, lanceolate, with a prominent midrib on their internal surface, rose-coloured, but towards the tips sanguineous. *Labellum* free, convolute, oblong-linear, truncate, emarginate. *Crest* spongy, yellowish, towards the base becoming more fleshy, and considerably elevated above the disk. At each side of its base, but seated on the disk, there is a small roundish yellow gland. Like the other segments of the perianth the labellum is rose-coloured, except its upper third, which is of the same colour as the tips of the petals. *Column* clavate, semiterete, white, the upper part of its internal face of a papillose nature and yellowish. *Stigma* infundibuliform, its lateral margins toothed. *Anther* large, fleshy, terminal, operculiform, subbilobed, purple, suspended by a lobed process of the upper part of the back of the column. *Germen* sessile, fleshy, cylindrical, about 2 inches long.

This species is nearly allied to *C. rosea, Lindl.,* but is well distinguished by its truncate and emarginate, not acute, labellum, and by the crest being entire at and below the middle. I possess three other species from Brazil—one from the province of Goyaz, with long narrow leaves, my only specimen of which is too imperfect to be described, and the two following :

Cleistes *montana,* (Gardn. Herb. Bras. n. 5879.) labello sepalorum longitudine trilobo lobis lateralibus lanceolatis acutis intermedio rotundato crispo integro, lamella per medium integra apicem versus denticulata.—HAB. In moist bushy places near the summit of the Organ Mountains, Brazil.

Cleistes *Miersii,* (Gardn. MSS.), labello convoluto oblongo-lanceolato acuminato integerrimo margine undulato sepalorum longitudine, lamellis apicem versus lacerato-denticulatis.—HAB. At Tijuca, in the province of Rio de Janeiro, *John Miers, Esq., G. Gardner.*

Fig. 1. Labellum and column. *f.* 2. Base of the labellum. *f.* 3. Column. *f.* 4. Anther-case :—*magnified.*

TAB. CDLXXV.

LINDENIA ACUTIFLORA.

Corollæ limbi laciniis acutis tubo 6-7-ies brevioribus.
HAB. Mexico. Puente-nacional, province of Vera Cruz, *Linden.*
n. 358.

This may possibly be a mere variety of the *Lindenia rivalis*
figured in the following Plate; but in the specimens we have
received from Mr. Linden, as well another presented to us
by Mr. Harris, and gathered by Mr. Galeotti, the leaves are
very much smaller and more downy, and the divisions of the
limb of the corolla are nearly one third less and very much
more pointed, although the tube is very nearly of the same
length as in *L. rivalis. Bentham.*

Tab. CDLXXV.

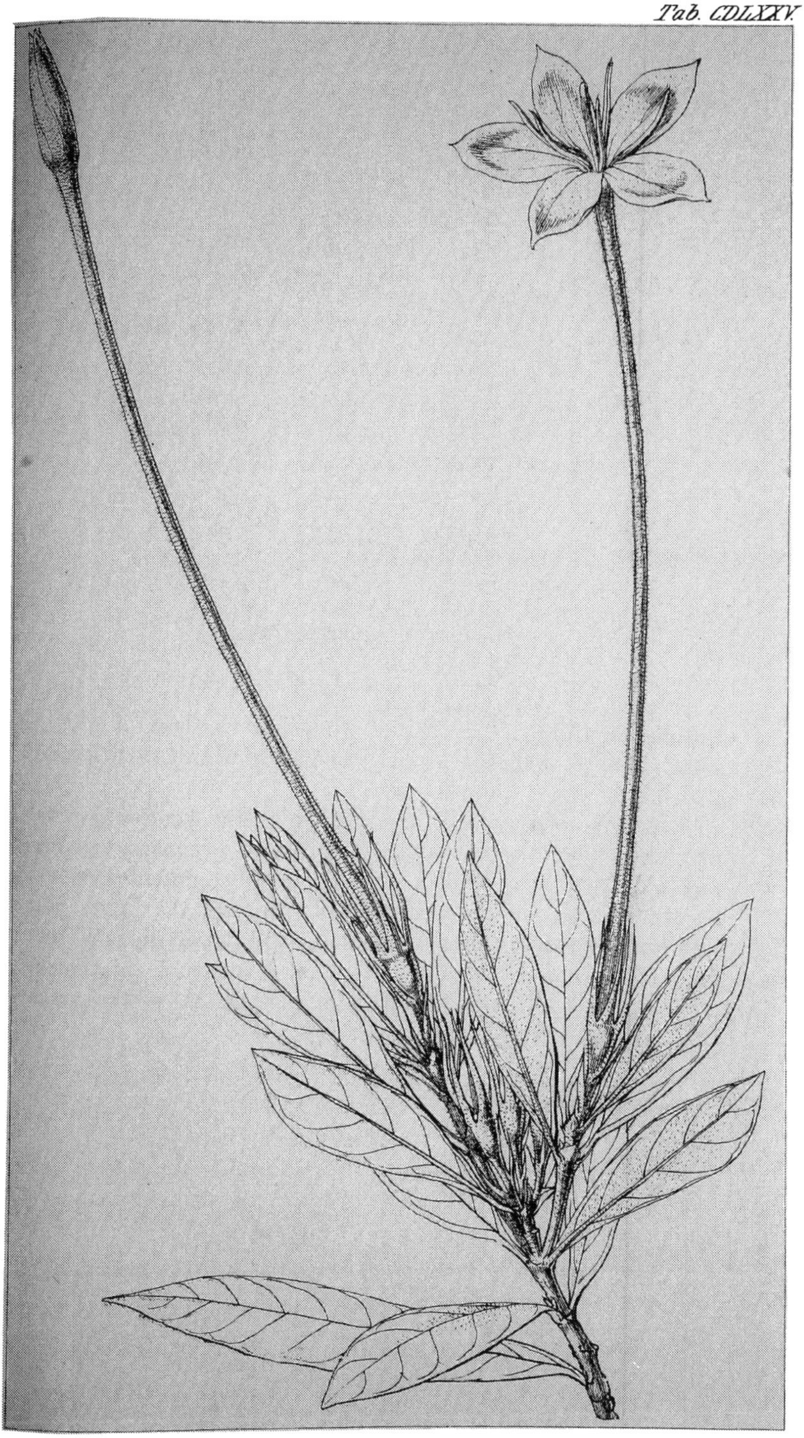

TAB. CDLXXVI.

Lindenia rivalis. *Benth.*

Gen. Char. *Calycis tubus* turbinatus, 5-costatus; *limbus* 5-partitus, laciniis angustis acutis. *Corolla* hypocrateriformis, tubo longissimo tenui æquali; limbo 5-partito, laciniis oblongis patentibus, æstivatione imbricatis. *Antheræ* 5, lineares, sessiles ad corollæ sinus. *Stylus* filiformis e basi glaber, apice incrassatus, brevissime bifidus, lobis intus stigmatiferis. *Capsula* laciniis calycinis coronata, bilocularis, placentis centralibus. Semina numerosissima angulata.—Frutices *Mexicani.* Folia *opposita, breviter petiolata, oblonga, ad apices ramorum conferta.* Stipulæ *utrinque solitariæ, fusco-membranaceæ, acuminatæ, in vaginam brevem connatæ, deciduæ.* Corymbus *terminalis condensatus pauciflorus.* Bracteæ *lineares.* Flores *subsessiles, albi.*

Lindenia *rivalis;* corollæ limbi laciniis obtusiusculis tubo 4-5-ies brevioribus. *Benth. Pl. Hartw.* 84.

Hab. Southern Mexico; on the banks of the river Teapa, *Linden, Herb. du Sud, n.* 331 : Guatemala; on the banks of streams, La Vera Paz, *Hartweg, n.* 581.

This is a shrub of two or three feet high, resembling in many respects the Brazilian genus *Augustea,* but differing in the form of the corolla, of which the tube is long, slender and straight as in *Tocoyena,* without the inflated oblique throat of *Augustea,* the style is also perfectly smooth. In this the original species the leaves become at length nearly smooth and attain the length of three or four inches. The tube of the corolla is about five inches* long, and the divisions of the limb rather more than an inch. *Bentham.*

* In Plantæ Hartwegianæ, by a mistake in copying, it is said to be 5 to 5½ *lines,* instead of *inches.*

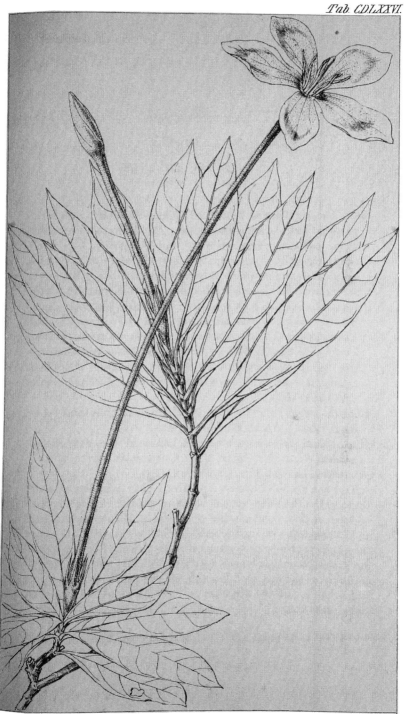

TAB. CDLXXVII.

COPTOPHYLLUM BUNIIFOLIUM. *Gard.*

GEN. CHAR. *Sporangia* ovata, vasculoso-reticulata, apice breviter radiatim striata, hinc longitudinaliter dehiscentia, biseriata, in laciniis frondis contractæ disposita. *Indusium*
nullum. *Sporulæ* subtriangulares, striatæ, glabræ.—Filiculæ
Brasilianæ, rhizoma *repente.* Frondes *cæspitosæ, dissimiles;*
sterilis *multifida, pinnulis linearibus dichotomis;* fertiles *tri-
pinnatæ, pinnulis sporangiferis, contractis;* venæ *furcatæ.—
Gard. in Hook. Lond. Journ. Bot.* 1, *p.* 133.

Coptophyllum *buniifolium;* glabrum, fronde sterili ovata multipartita, laciniis elongatis dichotomis, fertili laxe paniculata.
Gard. l. c.

Anemia dichotoma. *Gard. Herb. Bras. n.* 4084.

HAB. Among the *débris* of schistose rocks on the summit of
the Serra de Natividade, in the north of the province of
Goyaz, Brazil.

This and the following elegant little Fern, I have separated
from the genus *Anemia,* principally on account of their fertile
fronds rising distinctly from the rhizoma; and being in no way
connected with the stipes of the barren fronds. This latter
circumstance characterizes the true *Anemias,* for in them the
frond which bears the spikes of fructification is formed by the
union of two fertile fronds with one barren one. Since my
papers in the *Journal of Botany* were written, I have examined
the anatomical structure of the fertile frond of *Anemia Phylli-
tidis,* Sw., and I find that three nearly distinct bundles of
annular ducts can be traced to the top of the stipes, where they
at last separate, one running into the barren, and one into
each of the fertile portions. Link, I find, entertains similar
views on the structure of *Anemia.—G. Gardner.*

Fig. 1. Sporangium. *f.* 2. Sporules:—*magnified.*

1   2

TAB. CDLXXVIII.

Coptophyllum millefolium. *Gardn.*

Villosum, frondes terili oblonga vel ovato-oblonga multipartita, laciniis brevibus linearibus dichotomis, fertili elongata coarctata.

Coptophyllum millefolium. *Gard. in Hook. Lond. Journ. Bot. vol.* 1, *p.* 133.

Anemia millefolium. *Gard. Herb. Bras. n.* 4083.

Hab. Rare on dry arid hills near Villa de Arrayas, in the north of the province of Goyaz, Brazil.

No one, at first sight, would believe the barren fronds of the two plants which constitute the genus *Coptophyllum*, to belong to the tribe of *Ferns*, resembling, as they do, much more the leaves of some species of *Umbelliferæ*. The developement of parenchyma is here nearly reduced to its minimum, and consequently the dichotomous venation of the leaf is most beautifully and distinctly exhibited.—*G. Gardner.*

Fig. 1. Sporangium. *f.* 2. Sporules :—*all magnified.*

l

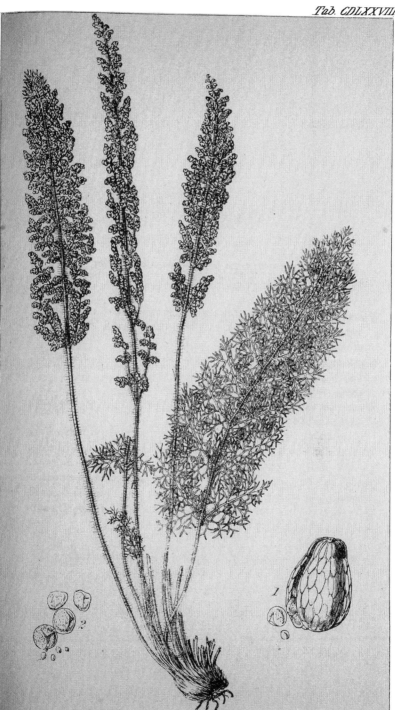

TAB. CDLXXIX.

IPOMÆA (STROPHIPOMÆA) GOYAZENSIS. *Gardn.*

Glaberrima, foliis late ovatis subtriangularibusve acutis basi profunde cordato-bilobis lobis approximatis, pedunculis trifloris, calycis glabri laciniis oblongis obtusis, corollæ tubo infundibuliformi limbo patente quinquelobo lobis emarginatis.
Ipomæa Goyazensis. *Gard. Herb. Bras. n.* 3909.
HAB. Rare, among bushes at the foot of the Serra de Santa Brida, province of Goyaz, Brazil.
Tota glaberrima. *Caules* teretes, volubiles. *Folia* alterna, petiolata, 3-5 poll. longa, 2-3½ poll. lata, majora subtriangularia, minora rotundato-ovata, apice acuta, basi profunde cordato-biloba, lobis approximatis, supra viridibus, subtus pallidioribus. *Petiolus* unciam longus, supra canaliculatus. *Pedunculi* axillares, breves, triflori. *Pedicelli* inæquales, intermedio longitudine circiter calycis, lateralibus brevioribus. *Calycis* foliola subæqualia, oblonga, obtusa, concava. *Corollæ* tubus albus, infundibuliformis, limbo violaceo, patente, 5-lobo, lobis emarginatis. *Stamina* erecta, tubo inclusa.
This very beautiful species of *Ipomæa* I only met with once, and then but sparingly in flower. It would be a most desirable object for cultivation, the tube of the corolla being pure white, while the limb is of a rich violet colour.—*G. Gardner.*

TAB. CDLXXX.

ACHIMENES RUPESTRIS. *Gardn.*

Suffruticosa, caule erecto glanduloso-villoso, foliis ternatim verticillatis breve petiolatis ovatis serratis acutis vel subacuminatis basi rotundatis utrinque glanduloso-pilosis, pedicellis axillaribus solitariis 1-floris, calycis 5-partiti lobis oblongis obtusis erectis, corollæ limbo amplo patente, lobis rotundatis. Achimenes rupestris. *Gardn. Herb. Bras. No.* 3874. HAB. In clefts of rocks near the summit of the Serra de Natividade, province of Goyaz, Brazil. Feb. 1840. *Suffrutex* pedalis, ubique glanduloso-pilosus. *Folia* ternatim verticillata, breve petiolata, 2 poll. longa, 12-16 lin. lata, ovata, serrata, acuta vel subacuminata, basi rotundata. *Calycis tubus* ovario adnatus; *limbus* 5-partitus, lobis oblongis, obtusis, erectis. *Corolla* pallide purpurea, tubo infundibuliformi, limbo amplo patente, 5-fido, lobis integris rotundatis. *Stamina* 4 didynama; antheris inter se cohærentibus.— G. Gardner.

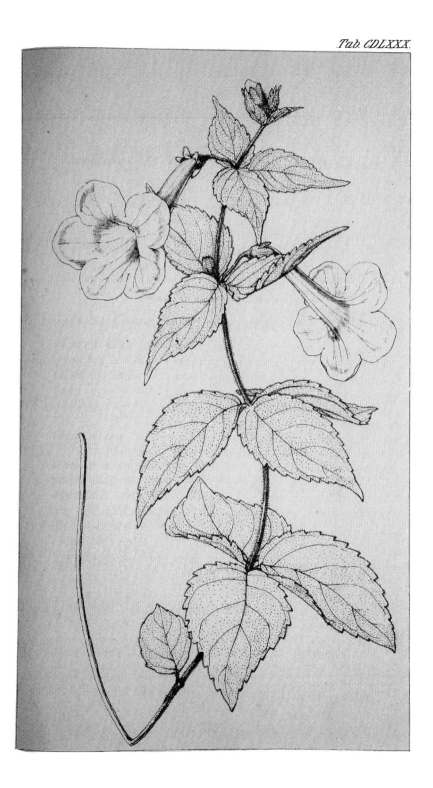

TAB. CDLXXXI.

Antidesma alnifolium.

Glabrum, foliis cordato-v. cuneato-rotundatis 3-5-nerviis grosse dentato-serratis, spicis axillaribus pilosis, masculis compositis, fœmineis simplicibus.

Hab. Eastern part of the Colony, Cape of Good Hope, *Mr. Bowie;* Port Natal, *Mr. Krauss, Herb. n.* 160.

It was my desire to give a name to a shrub that had been long cultivated at Kew, which induced me to figure and describe the present plant from very imperfect specimens, and of which only the *female* plant with immature ovaries, was known to me. This is a branching shrub, about 3 feet high, with the leaves variable, but not much unlike those of the Alder, the spikes of flowers axillary, scarcely longer than the petiole. The flowers clustered within small bracteas and sessile on the rachis. The perianth closely surrounding the germen, 5 to 6 cleft, the teeth obtuse. Germen ovate, styles 3 ; stigmas obtuse.—Long after the engraving was executed, I detected a *male* specimen of the same plant in Mr. Krauss' collection from Port Natal. Its spikes are long and compound, almost as long as the leaves. Flowers scattered, sessile. Perianth of 8 to 10 oblong segments, which are alternately smaller. Filaments 10 to 12, much exserted, with long hairs on the lower half. Anthers 2-celled, the lobes or cells rounded, spreading. Pistil none, but there are 3 or 4 fleshy glands at the base of the stamens. These male flowers are very small and not in perfect condition, being more or less eaten by insects ; so that this account of the fructification is necessarily incomplete, too much so to allow of my saying with certainty that it is an *Antidesma.* The ovary is so imperfect and minute, that it is difficult to distinguish its internal structure. It was believed to be 3-celled by the artist ; but the representation is probably erroneous.

Tab. CDLXXI. Female branch of *Antidesma alnifolium. Fig.* 1. Portion of a female spike. *f.* 2. Single flower. *f.* 3. The same with the perianth laid open. *f.* 4. Section of the ovary, but probably erroneously represented with 3 cells :—*magnified.*

Tab CCCLXXXI

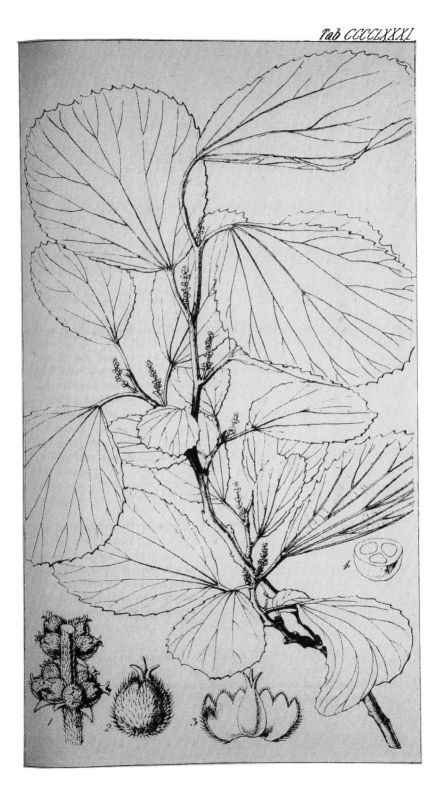

TAB. CDLXXXII.

CELASTRUS SUBSPICATUS.

Frutex glabra, ramis subverrucosis, foliis ovali-ellipticis acutis subcoriaceis serratis brevi-petiolatis, racemis compositis spicatis terminalibus rarissime axillaribus.

HAB. ———— ?

This is another plant, which like the *Antidesma alnifolium*, has been long cultivated in the Royal Botanical Gardens of Kew, and which flowers every summer, but of the history of which nothing is known; and it appears to be an undescribed species of *Celastrus*. The plant is 4 or 5 feet high, the branches flexuose and straggling, the leaves subcoriaceous, alternate, oval-elliptical, rather obscurely serrated, acute, paler and more conspicuously reticulated beneath; every where glabrous. The petioles are short, and in their axils are gemmæ with sharp, almost subulate, scales. The flowers are on short pedicels, and arranged in a compound mostly terminal spike or rather raceme, rarely axillary. Calyx cup-shaped, with 5 deep, rounded, obscurely denticulated lobes. Petals 5, obovate. Stamens 5, short, alternating with the petals, arising from a perigynous disc which lines the lower half of the calyx. Germen ovate, 3-celled, each with 2 ovules. Style short, thick. Stigmas 3, large, glandular.

Fig. 1. Flower from which the petals have been removed. *f.* 2. Entire flower. *f.* 3. Flower of which the calyx is laid open, and the petals removed. *f.* 4. Petal. *f.* 5. Vertical section of the pistil. *f.* 6. Transverse section of the germen:—*magnified.*

Tab CCCCLXXXII

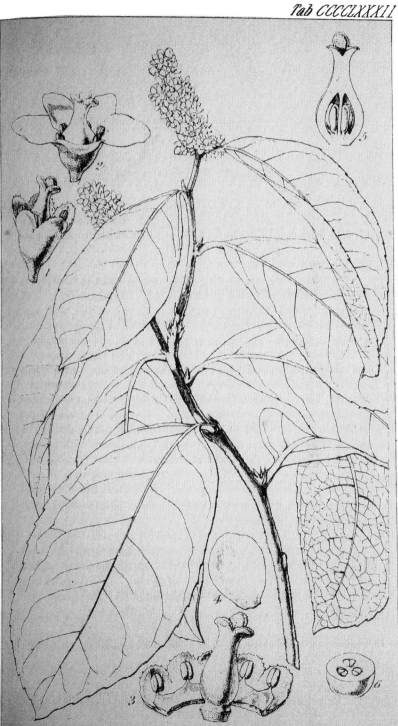

TAB. CDLXXXIII.

OXYRIA ELATIOR.

Caule aphyllo elato superne ramoso, racemis paniculatis, verticillis 6-12-floris, pedicellis fructiferis reflexis achenio subbrevioribus, sepalis interioribus obovato-subspathulatis obtusissimis, achenii suborbicularis alis membranaceis transverse venosis utrinque profunde cordato-incisis, foliis radicalibus longe petiolatis reniformibus margine obsolete crispato-undulatis. *Meisn.*

Oxyria elatior. *Brown in Wall. Cat. n.* 1726. *Meisn. in Wall. Plant. Asiat. Rar. v.* 3, *p.* 64.

HAB. The mountains of Emodi in Kamoun, (*Blinkworth.*) *Dr. Wallich.*

I am indebted for native specimens of this plant to Dr. Wallich; the living cultivated specimens, from which our figure is taken, were sent to me from the noble gardens of His Grace the Duke of Northumberland, at Syon House. Meisner well observes of it " *Oxyriæ reniformi,* Hook., nimis affinis, et vix differt, nisi statura altiore, sesquipedali, racemis longioribus magisque paniculatis, sepalis interioribus (i. e. erectis) apice dilatatis, obtusissimis, subtruncatis, paullo majoribus, fructus ala apice basique ad semen usque incisa (in *O. reniformi* autem subintegra v. basi tantum cordata.") It retains its characters in cultivation; yet I can hardly believe it distinct from our *O. reniformis* of Europe and of N. America. From the latter country, at Sitka, on the Pacific side, I possess specimens $2\frac{1}{2}$ feet high; and others nearly as tall from the Rocky Mountains.

Fig. 1. Cluster of flowers. *f.* 2. Single flower. *f.* 3. The same laid open. *f.* 4. Pistil. *f.* 5. Capsule:—*magnified.*

Tab CCCLXXXIII

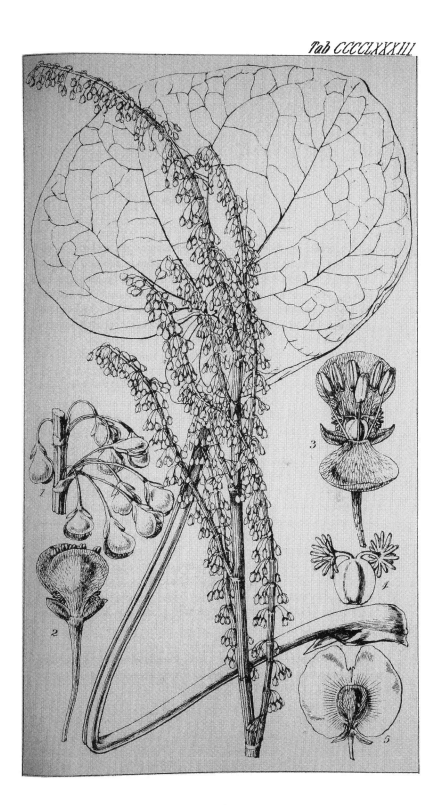

TAB. CDLXXXIV.

SAUVAGESIA DEFLEXIFOLIA. *Gardn.*

Fruticosa, caule erecto versus apicem ramoso, foliis deflexis lineari-lanceolatis marginatis subserratis acuminatis, stipulis subulatis setoso-pectinatis, pedicellis solitariis erectis, sepalis ovato-oblongis aristato-acuminatis supra medium subserrato-ciliatis, petalis obtusis.

Sauvagesia deflexifolia. *Gardn. Herb. Bras. n.* 3008.

HAB. Rare in a moist sandy upland campo near the mission of Duro, province of Goyaz, Brazil, Oct. 1839.

Caulis fruticosus, erectus, ad apicem ramosus, pedalis. *Folia* demum deflexa, alterna, vix petiolata, 4 lin. circiter longa, lineari-lanceolata, subserrata, acuminata, pellucido-marginata. *Stipulæ* subulatæ, setoso-pectinatæ, persistentes. *Flores* solitarii, axillares, pedunculati, pedunculis 2 lin. longis simplicibus, erectis, nunquam deflexis. *Calyx* quinquepartitus; sepalis ovato-oblongis, aristato-acuminatis, supra medium subserrato-ciliatis, margine scariosis. *Petala* 5, hypogyna, æqualia, obovata, obtusa, alba. *Stamina* hypogyna. *Staminodia* exteriora 10 spathulato-oblonga, interiora 5 petaloidea, petalis opposita. *Filamenta* fertilia 5, brevissima, staminodiis petaloideis basi adhærentia: *antheris* oblongis bilocularibus, loculis lateraliter dehiscentibus. *Ovarium* liberum, uniloculare. *Stylus* simplex. *Stigma* obtusum. *Capsula* ovato-oblonga, trivalvis. *Semina* plurima ad suturas valvarum biseriata.— *G. Gardner.*

Tab. CDLXXXIV. A fruit-bearing specimen of *Sauvagesia deflexifolia*. *Fig.* 1. Fruit. *f.* 2. Leaf and stipules. *f.* 3. Capsule, with the calyx removed :—*magnified.*

Tab. CCCCLXXXIV

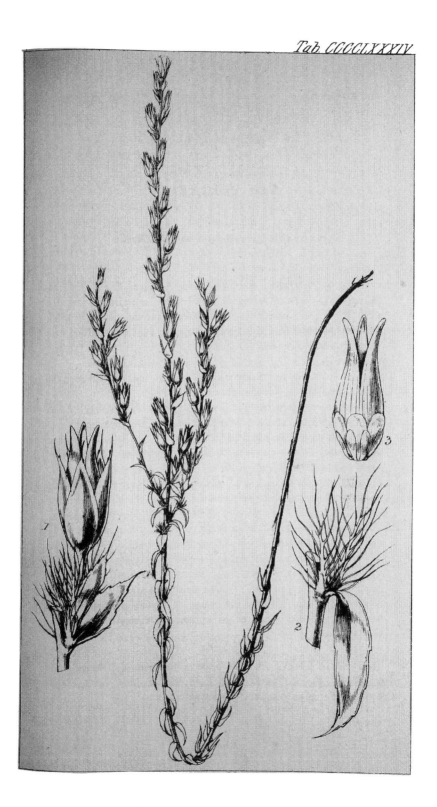

TAB. CDLXXXV.

CHILIOTRICHUM AMELLOIDES. *Cass.*

GEN. CHAR. *Capitulum* multiflorum, *fl. radii* ligulatis fœmineis uniseriatis, disci 5-dentatis hermaphr. *Invol. squamæ* imbricatæ oblongæ acutæ. *Recept.* convexum, paleis linearibus apice barbatis inter flores onustum. *Stigmata* fl. disci subulato-linearia elongata pubera. *Achænia* gracilia cylindracea angulato-striata. *Pappus* pluriserialis, setis filiformibus inæqualibus persistentibus.—Frutices *in extremâ Amer. Austr. spontanei parvi ramosi.* Folia *alterna sessilia coriacea integerrima, margine revoluta, supra glabra subtus plus minus tomentosa.* Pedunculi *solitarii* 1-*cephali tomentosi.* Ligulæ *albæ subtus purpurascentes.*

Chiliotrichum *amelloides;* foliis oblongo-ovatis basi angustatis planiusculis.

Chiliotrichum amelloides, *Cass. Dict.* 8, *p*, 576. De Cand. Prodr. 5, p. 216.

Amellus diffusus, *Forst. Comm. Goet.* 9, *p.* 39.

Tropidolepis diffusa, *Tausch, in Flora* 12, *p.* 68.

Aster Magellanicus, *Spreng. Syst.* 3, *p.* 526.

β. *lanceolatum,* (TAB. NOSTR. CDLXXXV.), foliis lanceolatis acutis basi attenuatis. *DeCand.*

γ? *rosmarinifolium,* Nees foliis linearibus intensius margine revolutis basi non angustatis.

Amellus rosmarinifolius. *Poepp. Exs.*

Ch. rosmarinifolium. *Less. in Linnæa,* 1831, *p.* 109. An species propria? (*De Cand.*)

HAB. Straits of Magellan, *Forster;* Cape Horn, Staten Land, *Dr. Eights,* (in Herb. Nostr.)—β. Falkland Islands, *D'Urville* and *Gaudichaud. Mr. Wright,* (in Herb. Nostr.)

This is perhaps one of the tallest shrubs in the Falkland Islands. Gaudichaud speaks of it as from 3 to 5 feet high, and Mr. Wright as 8 to 10 feet. The flowers are numerous, the ray pure white.

Fig. 1. Receptacle. *f.* 2. Floret of the disk with its scale. *f.* 3. Hairs of the pappus. *f.* 4. Floret of the ray:—*magnified.*

TAB. CDLXXXVI.

ASTER VAHLII. *Hook. et Arn.*

Herbaceus glaberrimus parce ramosus, foliis lineari-lanceolatis integerrimis obtusiusculis basi semiamplexantibus infimis spathulatis basi subvaginantibus subserratis, capitulis solitariis, involucri pauciserialis foliolis glaberrimis imbricatis linearibus acutis, radio purpureo, pappo cinereo, achenio villoso. Aster Vahlii. *Hook. and Arn. Contr. to Fl. of S. Am. in Hook. Comp. Bot. Mag. p.* 49. Erigeron Vahlii. *Gaudich. Fl. Isles Malouines, in Ann. des Sc. Nat. v.* 5, *p.* 103. *De Cand. Prodr.* 5, *p.* 295. HAB. Falkland Islands. *C. Darwin, Esq., Mr. Wright*; Cape Negro, Straits of Magellan, *C. Darwin, Esq.*; Andes of Chili, *Dr. Gillies*; Valdivia, *Mr. Bridges,* (*n.* 623); Chiloe, *Cuming,* (*n.* 55.)

Nearly allied to *Aster alpinus,* but at once distinguished by the glabrous leaves and stem, and involucre. Achenia sulcated, hairy.

Fig. 1. Floret of the disk. *f.* 2. Ditto of the ray. *f.* 3. Hairs of the pappus :—*magnified.*

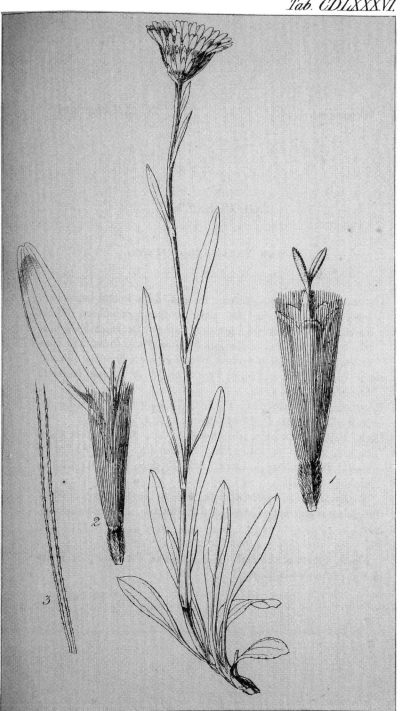

TAB. CDLXXXVII.

CHEILANTHES MONTICOLA. *Gardn.*

Frondibus pinnatis, pinnis oblongis obtusis crenatis glabris, basi superiore auriculatis.

Cheilanthes monticola. *Gardn. Herb. Bras. n.* 3557.

HAB. On the perpendicular face of Schistose rocks, in a deep narrow ravine near the summit of the Serra de Natividade, province of Goyaz, Brazil. January, 1840.

Radix fibrosa, fibrillis pilosis. *Rhizoma* parvum, subglobosum. *Frondes* plures, fasciculatæ, pinnatæ. *Stipes* vix pollicaris, atro-fuscus, hispidus, semiteres, supra canaliculatus. *Frons* 3-4 pollicaris. *Rachis* filiformis, glaber. *Pinnæ* alternæ, sessiles, 5 lin. longæ, oblongæ, obtusæ, crenatæ, glabræ, basi sursum auriculatæ, deorsum subtruncatæ. *Venæ* internæ, pinnatæ, furcatæ, venulisque divergentes, apice soriferæ. *Sori* marginales, oblongi. *Indusia* oblonga, membranacea, albida. *Sporangia* pedicellata, obovata, annulo fere completo cincta. *Sporulæ* subrotundæ, sub lente scabrellæ.

The only other species of *Cheilanthes* with simply pinnated fronds is *C. micropteris*, Sw., from the Andes of Peru. It differs from the present plant by its more slender habit and nearly rounded hairy pinnæ.—*G. Gardner.*

Fig. 1. Pinna. *f.* 2, 3. Sporangia. *f.* 4. Sporules :—*magnified.*

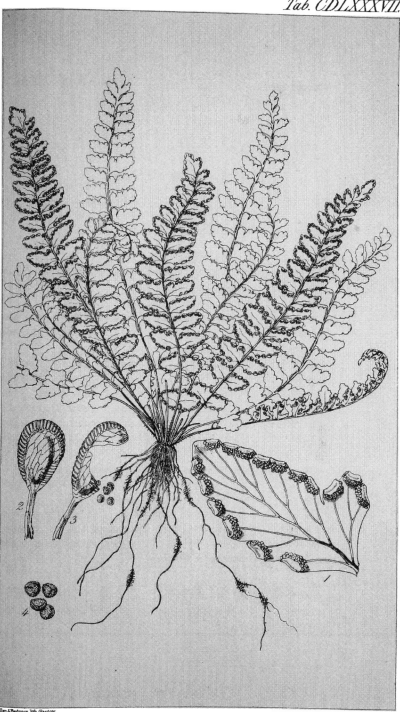

TAB. CDLXXXVIII.

Scolopendrium Lindeni.

Fronde lineari-oblonga obtuse attenuata integerrima basi cordata sublonge stipitata, stipite superne costâque inferne subtus ferrugineo-lanatis, venis ad basin bifurcatis venulis seu ramis liberis.

Hab. On old oaks, Chamulars, Prov. Chiapas, Mexico. *Linden, Herb. Mex. n.* 1543.

Caudex ——? *Stipes* subbiuncialis, superne squamis angustis subulatis ferrugineis lanosus. *Frons* vix spithamæa, ¾ unc. lata, lineari-oblonga, coriaceo-membranacea, integerrima, marginata, apice obtuse attenuata, basi cordata, superne glaberrima, subtus, præcipue ad costam basin versus, ferrugineo-lanosa. *Venæ* usque ad basin furcatæ; *venulæ* approximatæ, parallelæ, oblique horizontales, simplices, apicibus paulo infra marginem liberis clavatis; venula superiore et venula inferiore mox superioris venæ soriferis. *Sori* lineares longitudine variabiles.

This would be a true *Scolopendrium* of Presl, having the veinlets free at the apices, not there connected by reticulated veinlets as in his *Antigramma*. As a species it is quite different from any described one; but its nearest affinity is perhaps with *S. longifolium* of Presl (Reliq. Hænk. p. 48, t. 9, f. 1.), a native of Luzon. That, however, has much longer fronds, is quite glabrous, and tapers at the base into the stipes.

Fig. 1. Portion of the fructified frond. *f.* 2, 3. Sporangia. *f.* 4. Sporules. *f.* 5. Scale from the costa :—*magnified.*

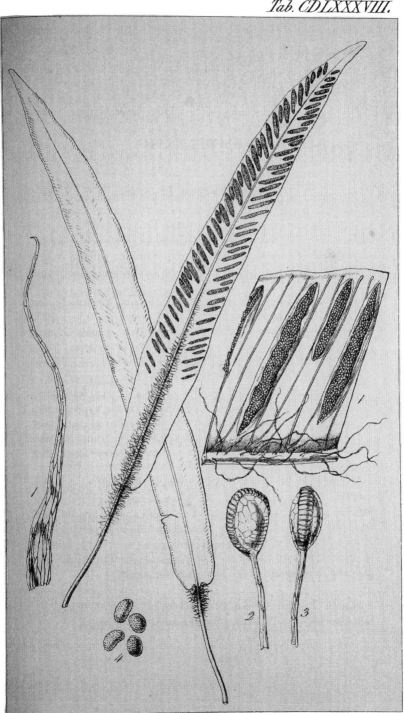

TABS. CDLXXXIX (AND CDXC.)

GUNNERA (Misandra) FALKLANDICA.

Dioica omnino apetala repens ferrugineo-hirsuta, pilis nunc deciduis, foliis reniformi-cordatis sublobatis crenatis petiolum subæquantibus, scapis folio brevioribus, floribus masc. et fœm. in spicam ovatam dense glomeratis, perianthiis glaberrimis.

Misandra Magellanica. *Gaud. in Ann. des Sc. Nat.* 5, *p.* 89.

(*vix* Gunnera Magellanica, *Lam. Dict. v.* 3, *p.* 61, *t.* 801, *f.* 2.)

HAB. Falkland Islands. *Gaudichaud, Mr. Wright.*

Whether or not I am correct in considering this distinct from *G. Magellanica* of the Straits of Magellan, must be left for more copious specimens and further observations to determine. I possess in my herbarium, from Valdivia, gathered by Mr. Bridges, (n. 647 of his herbarium,) what I consider to be the plant of Lamarck. It has leaves almost wholly glabrous; and petioles, even while the inflorescence is young, from 8 inches to a foot long, with the blade of the leaf shorter, broader, and exactly reniform, and the perianths (at least of the female, for I have not seen the male flower) very downy. Our present plant may also inhabit the Straits of Magellan as well as the Falkland Islands; for Mr. Bennett, in his valuable remarks on *Gunnera,*[*] observes, " Of *Misandra, two* species (both collected by Sir Joseph Banks and Dr. Solander) inhabit the dreary mountains of Tierra del Fuego; only one of these has yet been published." There exists, as is well known, so great a similarity between the vegetation of the Falkland Islands and that of Tierra del Fuego, that it is probable ours may be the 2nd species alluded to by Mr. Bennett. In the youngest of my flowering specimens I have been able to detect no petals to the male flowers; nevertheless I do not hesitate in thinking with Mr. Bennett and Endlicher that *Misandra* cannot generically be distinguished from *Gunnera.* The fruit is bright red and fleshy, each containing a small compressed stone. But the exact structure of the seed and embryo I have not been able to observe.

TAB. CDLXXXIX. *Fig.* 1. Male Plant; *nat. size. f.* 2. Small branch of male flowers; *magnified. f.* 3. Female plant; *nat. size. f.* 4. Female flower; *magnified.*

* Plantæ Javanicæ rariores, p. 74.

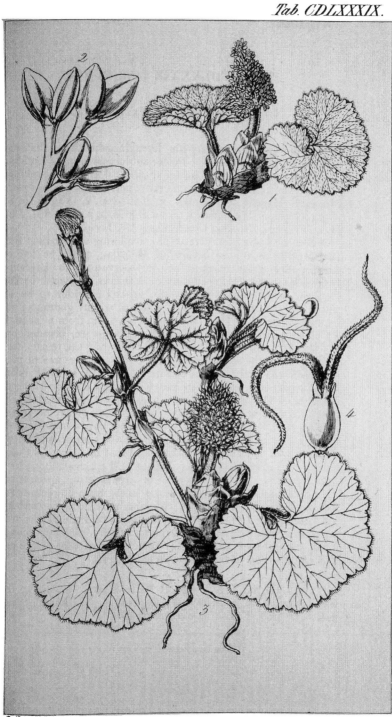

TAB. CDXC.

GUNNERA (Misandra) FALKLANDICA.

(Female Plant, noticed in the previous description.)

TAB. CDXC. Female Plant, with nearly mature fruit.
Fig. 1. Single fruit, with the dark purple styles still remaining.
f. 2. A drupe, cut through vertically. *f.* 3. Stone of the drupe:
—*magnified.*

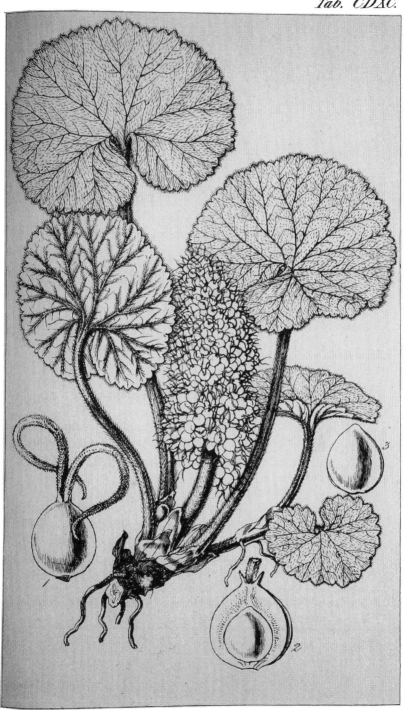

TAB. CDXCI.

HOMOIANTHUS ECHINULATUS. *Cass.*

Frutescens ramosus, caule ascendente tereti glabro dense folioso, foliis coriaceis basi dilatata semiamplexicaulibus linearibus subrecurvis siccitate transversim rugosis echinulatis, pedunculo terminali solitario folioso monocephalo, involucri squamis subtriserialibus oblongo-linearibus, ext. spinuloso-ciliatis, int. margine membranaceis.

Homoianthus echinulatus. *Cass. Dict.* 38, *p.* 458. *De Cand. Prodr.* 7, *p.* 65.

Perdicium recurvatum. *Vahl. Act. Soc. H. N. Hafn.* 1, *p.* 13, *t.* 7. *Gaudich.· in Ann. des Sc. Nat. v.* 5, *p.* 105. (*not Don, nor Poep.*)

Chætanthera recurvata. *Spreng. Syst.* 3, *p.* 503.

Perezia recurvata. *Less. in Linnæa,* 1830, *p.* 21. *Less. Syn. p.* 412. *Hook. et Arn. in Comp. to Bot. Mag. v.* 2, *p.* 42.

HAB. Falkland Islands. *Gaudichaud, Mr. Wright;* Straits of Magellan, *Commerson, Mr. Darwin, Capt. King.*

One of the most beautiful plants of the Falkland Islands, growing in peaty soil among rocks in large tufts, copiously branched, the branches bearing large, bright blue, very fragrant, flowers. The leaves, at least in a dry state, are singularly transversely wrinkled, and have their margins bent back so as almost to meet at the midrib.

Fig. 1. Leaf. *f.* 2. Involucre cut through, showing the receptacle. *f.* 3. Floret. *f.* 4. Hair from the pappus. *f.* 5. Tubular portion of the same, laid open to show the stamens and style. *f.* 6. Apex of style and stigmas :—*magnified.*

TAB. CDXCII.

BOLAX GLEBARIA. *Comm.*

Bolax Glebaria. *Comm.—Gaudich. in Ann. des Sc. Nat.* 5, *p.* 104, *t.* 3. *f.* 2. *De Cand. Prodr.* 4, *p.* 78.
Hydrocotyle gummifera. *Lam. Dict.* 3, *p.* 156, *t.* 189, *f.* 2 ? *(fig. mala.)*
Bolax gummifer et complicatus. *Spreng. Syst. Veget. vol.* 1, *p.* 879.
Azorella cæspitosa. *Vahl, Symb.* 348.
HAB. Straits of Magellan. *Commerson;* Falkland Islands, *Gaudichaud, Mr. Wright*; Chili and Patagonia, (*De Candolle.*)

Among some interesting drawings of Falkland Island scenery, brought home by Mr. Wright, a remarkable feature in the country is due to the frequent occurrence of the singular rounded and very compact tufts of this little Umbelliferous plant. "What can be more surprizing, in speaking of the vegetation of these islands," says D'Urville, "than the enormous tufts of *Bolax?* At first, their form resembles small mole-hills covered with green turf; but annually their young shoots, continually renewed, augment their original dimensions, till at length the mass attains a diameter and a height of many feet!" "It is an umbelliferous plant," he continues, "almost microscopic, respecting the nature of which the most experienced eye is apt to be deceived, so much is its appearance at variance with the usual aspect of the family to which it belongs. A resinous substance, of most powerful odour, exudes from every part of the plant and announces its presence from a distance." Mr. Wright speaks of the tufts being so large as to resemble small hay-stacks. The root is very long and tapering, perennial; branches numerous, dichotomous, proliferous. I find two varieties; the one which alone I have seen with fructification, has excessively dense and small foliage (f. 4-8) and is quite glabrous on the outside of the leaves, stellato-pubescent within; the other (f. 1-3) has larger leaves, stellato-pubescent on both sides; all are trifid, concave, with large concave sheathing glossy bases. Umbels sessile, 3-4-flowered. Leaves of the involucre ovate, entire, with sheathing bases (f. 6.)

Fig. 1. Larger *var.* of *Bolax Glebaria, nat. size. f.* 2, 3. Leaves, *magnified. f.* 4. Tuft of the smaller *var.* with scarcely mature fruit, *nat. size. f.* 5. Leaf. *f.* 6. Umbel. *f.* 7. Fruit. *f.* 8. Transverse section of ditto, *magnified.*

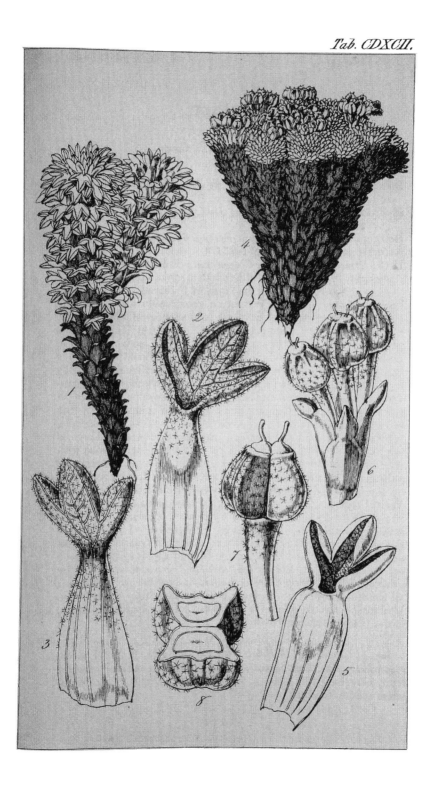

TAB. CDXCIII.

SENECIO LITTORALIS. *Gaudich.*

Caule erecto superne præcipue ramoso tereti, foliis sessilibus
lineari-lanceolatis pungenti-acutis marginibus revolutis basi
dilatatis semiamplexicaulibus, ramis foliosis monocephalis,
capitulis radiatis, involucri calyculati squamis circiter 20
lineari-lanceolatis disco æqualibus, floribus disci 30-40, ligulis
12-14, receptaculo convexo nudo.

Senecio littoralis. *Gaudich. in Ann. des Sc. Nat.* 5, *p.* 104. *De
Cand. Prodr.* 6, *p.* 413.

a. *lanatus*; foliis ramisque albo-tomentosis. *Gaudich. l. c.*

β. *glabratus*; foliis glabris. *Gaudich. l. c.* (TAB. NOST. CDXCIII.)

HAB. Falkland Islands. *Gaudichaud, Mr. Wright.*

The specimen here represented has leaves glabrous or nearly
so; but about the branches and particularly about the leaves is
an arachnoid wool, which looks as if, in the early state, it had
covered the whole of the stem, branches and involucre; the a.
of Gaudichaud is woolly all over. Perhaps *S. vaginatus,* Hook.
and Arn. Fl. of S. Am. in Journ. of Bot. v. 3, p. 331, should be
referred to this species.

Fig. 1, 2. Upper and under side of a leaf. *f.* 3. Involucre cut
through to show the receptacle. *f.* 4. Floret of the disk.
f. 5. Hair from the pappus. *f.* 6. Floret of the ray :—*magnified.*

TAB. CDXCIV.

OXALIS ENNEAPHYLLA. *Cav.*

Acaulis, radice bulbifera squamosa, petiolis longissimis, foliolis 9-20 cuneato-oblongis profunde bilobis subpilosis obtusis, scapis unifloris petiolo longioribus sub florem bibracteatis, sepalis oblongis villosis apice bipunctatis, staminibus longioribus stylos hirsutos superantibus.

Oxalis enneaphylla. *Cav. Ic. v.* 5, *p.* 7, *t.* 411. *De Candolle Prodr.* 1, *p.* 702. *Gaudich. in Ann. des. Sc. Nat.* 5, *p.* 105.

β. *pumila*; minor magisque pilosa.

Oxalis pumila. *Gaudich. in Freyc. Voy.* 1, *p.* 137.

HAB. Falkland Islands. *Née, Gaudichaud, Mr. Wright.*

This must be a very handsome plant, with its copious foliage and large showy white flowers. Its acid property is well known in its native country. Pernetty calls the plant " *Vinaigrette*," and Mr. Wright speaks of it as eaten in pies, and used instead of apple sauce.

Fig. 1. Calyx with stamens and pistil. *f.* 2. Stamens and pistil. *f.* 3. Pistil:—*magnified.*

n

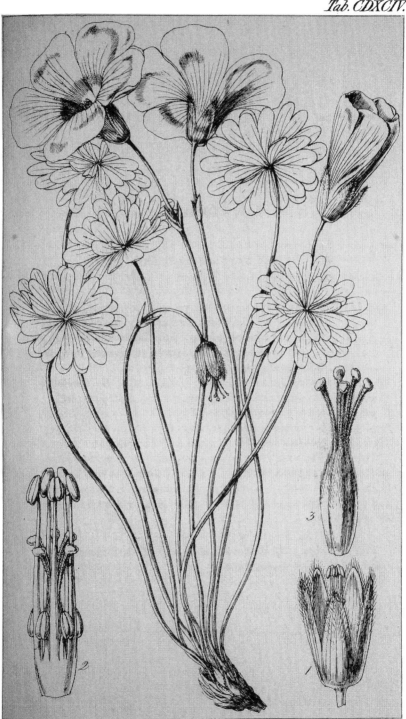

TAB. CDXCV.

RUBUS GEOIDES. *Sm.*

Caulibus repentibus petiolisque filiformibus, foliis trisectis lobo
  terminali maximo ovato obtuso irregulariter serrato lateralibus
  minimis sæpe nullis aut cum terminali coalitis, pedunculis
  solitariis unifloris petiolo multo brevioribus.
Rubus geoides. *Sm. Ic. Ined. t.* 19.
Dalibarda geoides. *Pers.—De Cand. Prodr.* 2, *p.* 568.
HAB. Straits of Magellan. *Commerson;* Falkland Islands, *Gau-*
  dichaud, Mr. Wright.
Sir James Smith was only acquainted with the flowering state
of this plant. My specimens from Mr. Wright are in fruit,
but they confirm the views of Sir James Smith respecting its
genus; for it is entirely the fruit of a *Raspberry*, being very
juicy, transparent, and delicious to the taste. The flavor Mr.
Wright describes as between a Raspberry and Strawberry. Our
flowering specimen is copied from Sir James Smith's figure, in
order that our representation may be the more complete.

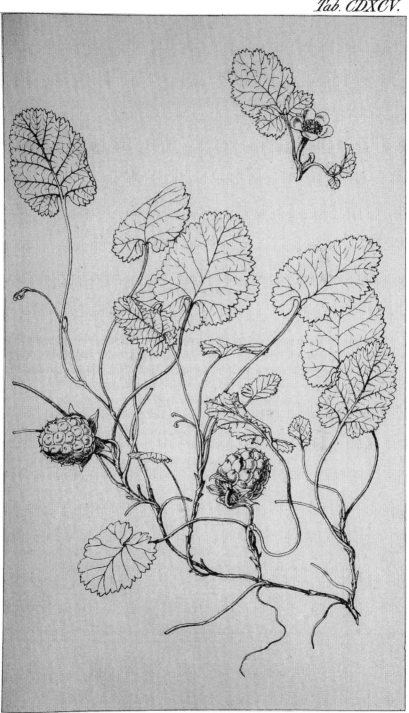

TAB. CDXCVI.

CHABRÆA SUAVEOLENS. *DC.*

Tota præsertim ad apicem lanuginosa, caule simplici folioso monocephalo, foliis radicalibus oblongis obtusis pinnatifidis sensim in petiolum longum attenuatis lobis approximatis rotundatis caulinis semiamplexicaulibus lanceolatis acuminatis superioribus integerrimis, involucri squamis lineari-oblongis lana immersis, stam. filamentis superne glanduliformibus.

Chabræa suaveolens. *De Cand. Prodr.* 7, *p.* 59.

Perdicium suaveolens. *Gaudich. in Freyc. Voy. p.* 125.

Lasiorhiza ceterachifolia. *Cass. Dict.* 43, *p.* 80. *Less. in Linnæa,* 1830, *p.* 11.

Lasiorhiza viscosa. *Cass. Dict.* 43, *p.* 80 ?

HAB. Falkland Islands. *Née, Gaudichaud, Mr. Wright.*

A very handsome showy species with large and highly fragrant flowers, which some authors, as Pernetty, compare to the odour of Benzoin, and others (Gaudichaud) to that of Vanilla.

Fig. 1. Involucre cut through to show the receptacle. *f.* 2, 3. Florets. *f.* 4. Stamens. *f.* 5. Hair of the pappus :—*magnified.*

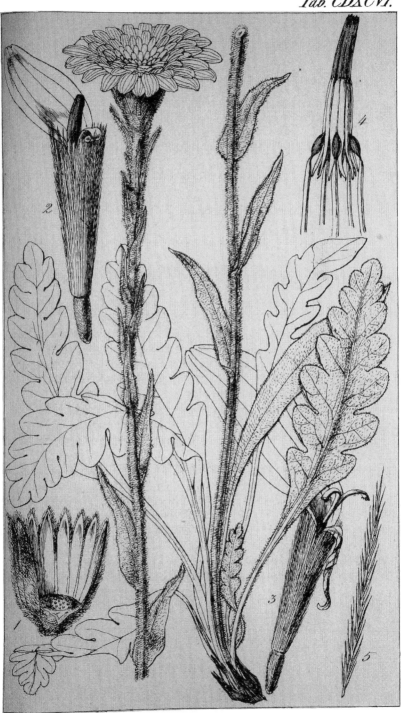

TAB. CDXCVII.

RANUNCULUS BITERNATUS. *Sm.*

Caule repente, foliis longe petiolatis circumscriptione cordatis 3-partitis partitionibus petiolulatis iterum sæpe tripartitis lobis cuneatis 3-fidis, petalis 6-8 oblongis (flavis), carpellis plurimis ovatis compressis stylo recurvato mucronatis in globum digestis.

Ranunculus biternatus. *Sm. in Rees, Cycl. n.* 48. *De Cand. Prodr. v.* 1, *p.* 30. *Deless. Ic. Sel. t.* 24.

HAB. Straits of Magellan. *Commerson*; Falkland Islands, *Mr. Wright.*

This species does not appear to have been discovered in the Falkland Islands till Mr. Wright detected it there. I should not have figured it, had I ascertained, before the engraving was prepared, that it was the same with the *R. biternatus* already so well represented by Delessert in his valuable *Icones.* I was misled by De Candolle's placing this plant in his division of the genus "*floribus albis*," whereas the inflorescence is decidedly yellow, as indeed Sir James Smith had suspected, whose description is moreover very accurate.

Fig. 1. Flower. *f.* 2. Head of carpels. *f.* 3. Single carpel :— *magnified.*

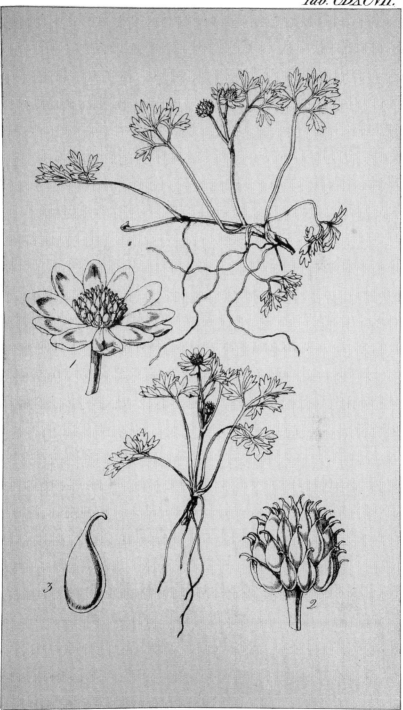

TAB. CDXCVIII.

ARABIS MACLOVIANA.

Glaberrima inferne ramosa, ramis erectis teretibus, foliis sub-
glaucis inferne dentato-serratis radicalibus ovato-oblongis
longe petiolatis caulinis sensim minoribus brevi-petiolatis
supremis lineari-oblongis sessilibus, corymbis compactis,
calycibus patenti-hirsutis pedicellos superantibus, petalis
spathulatis albis, siliquis erectis strictis uncialibus sublato-
linearibus stylo breviter terminatis, valvarum nervibus 3
crassis et reticulatim venosis.

Brassica Macloviana. *Gaudich. in Freyc. Voy.* 1, *p.* 137.

HAB. Falkland Islands. *Gaudichaud, Mr. Wright.*

M. Gaudichaud's description of *Brassica Macloviana* so entirely accords with this plant, that I have no hesitation in pro-
nouncing the two to be the same; but I do not see that the species can be referred to *Brassica.* It possesses quite the habit and
I think the character of *Arabis.* The valves of the siliqua have
three peculiarly strong prominent nerves, and the lateral ones,
being perhaps the most prominent, give a somewhat 4-angled
or 4-sided character to the fruit.

Fig. 1, 2. Flowers. *f.* 3. Stamens and Pistil. *f.* 4. Petal.
f. 5. Fruit. *f.* 6. Portion of a valve of the siliqua. *f.* 7. Seed.
f. 8. The same laid open. *f.* 9, 10. Embryos:—*magnified.*

TAB. CDXCIX.

Viola maculata. *Cav.*

Stigmate apice subplano, rostro brevissimo, caule abbreviato, foliis ovatis subtus sæpe minute fusco-punctatis longe petiolatis serratis puberulis, stipulis ovatis fimbriato-ciliatis, calcare brevi obtuso, petalis barbatis.

Viola maculata. *Cav. Ic. v. 6, t. 539. De Cand. Prodr.* 1, *p.* 297. *Hook. and Arn. in Bot. Misc. v.* 3, *p.* 144, *and in Bot. of Beech. Voy. p.* 10.

V. pyrolæfolia. *Poir. Dict.* 8, *p.* 836. *Gaudich. in Ann. des Sc. Nat.* 5, *p.* 104. (*excl. Syn.* V. Magellanicæ, *Forst.*)

V. lutea, foliis non acutis. *Feuill. Chil.* 3, *p.* 66, *t.* 48.

Hab. Falkland Islands. *Née, Gaudichaud, Mr. Wright*; Straits of Magellan to Conception in Chili. *Messrs. Lay and Collie, Cuming, &c.*

This inhabits the sands and sea-shores of the Falkland Islands, and, probably, similar localities in the Straits of Magellan and in Chili. The name is very inappropriate; the minute dots on the underside of the leaves which gave origin to it being almost microscopic, and not always present. The flowers are yellow, and no doubt can exist of the plant being the " *Viola lutea, foliis non acutis*," (for the leaves are frequently obtuse,) of *Feuill. Chil.*

Fig. 1. Stamens and pistil. *f.* 2. Single anther. *f.* 3. Pistil. *f.* 4. Petal :—*magnified.*

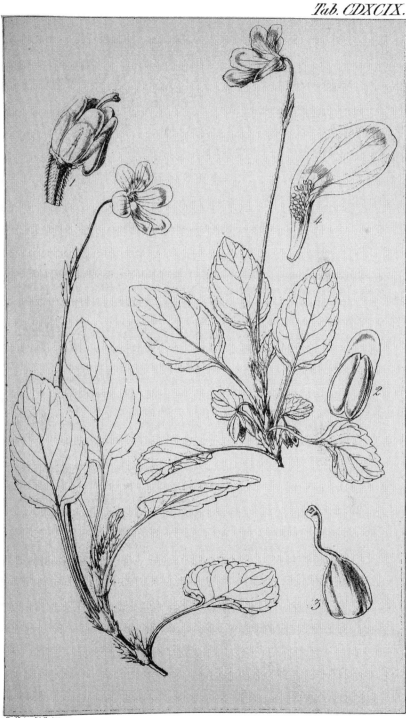

TAB. D.

ARACHIS MARGINATA. *Gardn.*

Perennis, caule brevi subsimplici sericeo-villoso, stipulis folio-
lorum par infimum superantibus, foliolis coriaceis obovatis
oblongisve emarginatis supra glaberrimis subtus sericeo-vil-
losis margine valde incrassatis longe ciliatis.

Arachis marginata, *Gardn. Herb. Bras. n.* 3103.

HAB. Rare in upland sandy Campos near the mission of Duro,
province of Goyaz, Brazil.

The specimen from which the figure has been taken is
perhaps only the young state of a much larger plant, as all the
other species of the genus have long, procumbent, lateral
branches. It differs from *A. villosa*, Benth., by the leaves
being smooth above; from *A. tuberosa*, Bong., by the long silky
hairs which cover the whole plant, except the upper surfaces of
the leaflets, and the much less reticulated foliage; and from
both by this latter being more coriaceous and having a much
thicker margin. My *n.* 2091, from Piauhy, is a broad-leaved
form of *Arachis pusilla*, Benth. in Trans. Linn. Society, 18, p.
159.— *G. Gardner.*